Derrida after the End of Writing

John D. Caputo, *series editor*

PERSPECTIVES IN
CONTINENTAL
PHILOSOPHY

CLAYTON CROCKETT

Derrida after the End of Writing
Political Theology
and New Materialism

FORDHAM UNIVERSITY PRESS
New York ■ 2018

Fordham University Press has no responsibility for the persistence or accuracy of URLs for external or third-party Internet websites referred to in this publication and does not guarantee that any content on such websites is, or will remain, accurate or appropriate.

Fordham University Press also publishes its books in a variety of electronic formats. Some content that appears in print may not be available in electronic books.

Visit us online at www.fordhampress.com.

Library of Congress Cataloging-in-Publication Data

Names: Crockett, Clayton, 1969– author.
Title: Derrida after the end of writing : political theology and new materialism / Clayton Crockett.
Description: First edition. | New York, NY : Fordham University Press, 2017. | Series: Perspectives in Continental philosophy | Includes bibliographical references and index.
Identifiers: LCCN 2017003972 | ISBN 9780823277834 (cloth : alk. paper) | ISBN 9780823277841 (pbk. : alk. paper)
Subjects: LCSH: Derrida, Jacques. | Religion. | Political science—Philosophy.
Classification: LCC B2430.D484 C76 2017 | DDC 194—dc23
LC record available at https://lccn.loc.gov/2017003972

Printed in the United States of America

20 19 18   5 4 3 2 1

First edition

for Jack, for Jacques and Everything After

Contents

Derrida after the End of Writing

Introduction
Derrida and the New Materialism

This is a book about Jacques Derrida. In it I try to open up some new ways to read his philosophy by focusing on his emphasis on religion and politics toward the end of his career, and then using this to develop a more materialist reading of Derrida. I have had a lot of inspiration for this project, including the work of John D. Caputo, Catherine Malabou, and Karen Barad. My contention is that Derrida's thought remains important; it cannot be relegated to the dust-bin of some late–twentieth-century linguistic idealism and subjectivist constructivism that just plays with language. This has always been the wrong understanding of Derrida, from its earliest incarnation, but this bad reading has been reasserted with some of the newer theoretical currents in the twenty-first, such as Speculative Realism, New Materialism, and Object-Oriented-Ontology, not to mention the political Lacanianism of Slavoj Žižek and Alain Badiou.

Here I do claim that something changes in Derrida's work, but this shift cannot be described precisely as a "turn." Using Malabou's idea of a motor scheme, an organizing image of thought or root metaphor that expresses the broadest information of a time period or epoch, I suggest that Derrida's philosophy works mostly within the motor scheme of writing. At a certain point, however, during the late 1980s and early 1990s, this cultural-intellectual-technological scheme of writing evolves into a motor scheme that Malabou describes as one of plasticity. For Derrida, the works of the 1990s and early 2000s are different because they are written in some ways as a response to and expression of this change of scheme. In effect, the concerns of ethics, politics, and religion emerge into the foreground as writing becomes more and more backgrounded. Michael Naas gets at the heart of what animates Derrida's work with the title of his important book, *Miracle and Machine*, because Derrida is explicitly more engaged with the tension between belief and responsibility as a kind of singular, miraculous living event and the repetition of a kind of machinic technicity that exposes this life to a form of death from the beginning.[1]

We struggle to read Derrida because something is different and we are not sure what it is. I think that Malabou's notion of a motor scheme and her distinction between writing and plasticity as motor schemes are important heuristics for helping us understand Derrida. I am less invested in as sharp a distinction as Malabou makes with this change in scheme, but I am using it to explore what it would mean to read Derrida beyond the scheme of writing. Malabou collapses the two "sides" of Derrida's later work, the responsibility of the living being and the mechanical repetition that gives death, into her understanding of neurobiological form, which is characterized by plasticity. For his part, Derrida increasingly adopts a biological metaphor of auto-immunity to make sense of religion and politics, most significantly in his explicit essay on religion, "Faith and Knowledge." This notion of auto-immunity exceeds any simple or even extended sense of writing.

I claim, then, that something changes in Derrida's later work that is not simply a "turn," but a more background context of and for his work, which is well articulated with Malabou's idea of a motor scheme. There is a kind of transition from an intellectual motor scheme based on writing in a broad sense to one based on what Derrida sometimes characterizes in terms of the machinic, teletechnology, or technoscience, and Malabou calls plasticity. Arthur Bradley calls this situation an "originary technicity" in his book on technology from Marx to Derrida.[2] This transformation in the 1980s and 1990s changes how Derrida writes and how he is read in his later work. Caputo is one of the first American readers to really appreciate this, although he presents his interpretation more in terms of religion than in terms of plasticity or technicity. But I think that many of the arguments about Derrida's engagements with religion and with politics in his later work are tied to this shift in one way or another. It's not that Derrida changes his philosophy; he is clear about how consistent his interests, ideas, and themes are across his career. Rather, something has changed in the background or the cultural and intellectual context of how we read him.

My interpretation, however, is not just about getting the correct exegesis of Derrida's works. It also consists of an intervention concerned with developing a constructive understanding of Derrida that shows his continuing relevance for contemporary philosophical discussions, including those concerning New Materialism, speculative realism, ideas about ecology and the natural sciences, and object-oriented ontology. That is, within a new intellectual context in the twenty-first century, we need new resources and new ways of seeing how his thought is important and relevant beyond simple polemics (whether pro or anti). Here is where I constructively engage Derrida with Jacques Lacan, to a certain extent, and also with Caputo's radical theology, with Malabou's biological materialism, and with Barad's understanding of quantum physics as a materialist hauntology.

To read and think about Derrida beyond the motor scheme of writing is to engage with the religious and political significance of his later work. I want to take this one step further and argue that working through these political and religious themes opens the possibility for a more materialist interpretation of Derrida. Derrida certainly kept a critical distance from materialism; he does not use this term in a positive sense. At the same time, I think that the non-reductionist materialism expressed in terms of New Materialism offers important tools to understand Derrida. In some ways, I am appropriating Derrida as a new materialist, but I don't think that deconstruction proscribes such an entanglement.

Here I will specify the arguments and themes of the specific chapters of the book, before returning later in the introduction to this topic of New Materialism. The first chapter, "Reading Derrida Reading Religion," is a more straightforward account of religion in Derrida's philosophy. In this chapter, I provide an introductory and background analysis to set the stage for the rest of the book. I discuss Derrida's work as a whole, although I do not engage it comprehensively or systematically. I argue that religion is a constant element of Derrida's work, even as the treatment of religion shifts in his later work. Chapter 2, "Surviving Christianity," focuses more specifically on the idea of the deconstruction of Christianity, including how Jean-Luc Nancy makes the deconstruction of Christianity the main theme of deconstruction in general, and how Derrida both endorses Nancy's project and keeps his distance from it. I suggest as a conclusion that it may be better to mourn Christianity while surviving it than to simply try to overcome it, which ironically perpetuates a triumphalist Christianity. I also consider Derrida's interpretation of Shakespeare's *Merchant of Venice*, which opens onto a discussion of Gil Anidjar's impressive book *Blood*.

In Chapter 3, "Political Theology Without Sovereignty," I engage Derrida's later work more directly, and place it in contact with the tradition of political theology. Much of Derrida's later philosophy involves a critique of the notion of sovereignty, which is understood in generally Schmittian terms. Insofar as political theology rotates around sovereignty, Derrida wants nothing to do with it. At the same time, if we think about political theology in distinction from sovereignty, then it is possible to read Derrida's thought in terms of a political theology without sovereignty. I suggest that much of Derrida's later philosophy consists of an implicit engagement with Schmitt, and offers ways to think about political theology that do not presuppose a Schmittian paradigm, partly by drawing on the work of Jeffrey W. Robbins.

Chapter 4, "Interrupting Heidegger," reconsiders the theme of sovereignty by way of an interruption, by considering how Derrida uses the poetry of Paul Celan to contest Heidegger. In particular, I focus on Derrida's essay "Rams," where he analyzes a poem by Celan, "Vast Glowing Vault," as a way to call into question Heidegger's three theses concerning the concept of world from his lecture course on *The Fundamental Concepts of Metaphysics*. For Heidegger, the stone is without world, the animal is poor in world, while the human being or Dasein is characterized as world-building. Derrida quotes Celan's final line, "The world is gone / I must carry you" as a way to challenge Heidegger's theses. So many of Derrida's later reflections return to his complication and contestation of these Heideggerian theses, including his engagements with animality. Finally, at the end of his last seminar on *The Beast and the Sovereign*, Derrida returns to Heidegger and considers the sovereignty of death as something that both supersedes and explodes our ideas about human sovereignty.

Lacan is an important resource throughout this book, and I claim that Derrida is an extremely careful as well as critical reader of Lacan. In Chapter 5, "Derrida, Lacan, and Object-Oriented Ontology," I use Lacan's thought to bring Derrida into contact with Speculative Realism and Object-Oriented Ontology as a way to reflect on philosophy of religion "at the end of the world." The promise of recent philosophies of SR and OOO is their engagement with the physical sciences, and their understanding of objects as extremely complex entities with their own properties and qualities apart from human conscious perception and language. At the same time, some of these philosophies go too far in their attempt to eliminate human subjectivism, and in their critiques of poststructuralism. I think it is more interesting to read philosophers such as Derrida, Lacan, and Deleuze creatively, in more speculative and object-oriented ways, than to read OOO over against poststructuralism. In this chapter, I discuss Timothy Morton's book and idea of *Hyperobjects*, and use this conception to think about philosophy of religion at the end of the world. For Morton, a hyperobject is a very strange object, and it helps us think about how Lacan's object *petit a* is also an object, and how we can imagine OOO as aOO (*a*-oriented ontology). Furthermore, for Derrida, and picking up on the theme of objects in the world introduced in Chapter 4, we can pay more attention to the third of Heidegger's objects, the stone or the natural object, which Heidegger says completely lacks a world. Derrida contests all of Heidegger's theses, and he spends a lot of time specifically on the theme of the animal in his later work, but Derrida does not really elaborate on this specific thesis of the stone. OOO offers resources to draw out this underdeveloped theme of Derrida, the rejection of how Heidegger characterizes the object, a stone, and the possibility for expanding the context and horizon in which we read and think about Derrida's philosophy.

After this critical and constructive engagement with OOO, I then turn in three consecutive chapters to explicitly treat the (for me) three most significant contemporary interpreters of Derrida, each of whom takes Derrida's philosophy to new places and new contexts that are in some sense beyond writing (as is OOO). Caputo is the most important contemporary philosopher of religion in the United States, and after his influential book on *The Prayers and Tears of Jacques Derrida*, he has pivoted to take up the challenge of radical theology. Although Caputo's interpretation of Derrida has become a standard and even stereotypical understanding of the religious aspects of Derrida's philosophy, against which newer readings of Derrida, such as Martin Hägglünd's argue, I claim that Caputo's interpretation is more complex than it may sometimes appear. Caputo radicalizes a certain reading of Derrida, not against Derrida, but in a profound affirmation of a religion without religion, and this attends to something that does change in Derrida's later work. Caputo uses his sophisticated understanding of Derrida's philosophy to develop his own, radical Derridean theology, based on the concept of the event in *The Weakness of God*, and then the notion of "perhaps" in *The Insistence of God*.

From Caputo I turn to the philosophy of Catherine Malabou. In many ways, Malabou provides the framework for my understanding of Derrida, with her distinction between a motor scheme of writing and a motor scheme of plasticity. Furthermore, I think that she is perhaps the most brilliant and creative contemporary philosopher in her own right, and she takes Derrida's philosophy in important and unforeseen directions. Chapter 7, "Deconstructive Plasticity," traces the work of Malabou beyond Derrida as she develops a biological materialism that is informed by deconstruction even as it transforms our understanding of deconstruction. Malabou's signature term is "plasticity," and she engages with brain plasticity and later with biological evolution and epigenetics in her work in important ways. Derrida's terms include mechanics, automaticity, tele-technology, and techno-science, and the two elements of this generalized technicity are what Michael Naas calls miracle and machine, which can be distinguished but cannot be fully separated. Malabou collapses both of these elements into her conception of form, and the inherent plasticity of form includes its own "miracle" of auto-annihilation or destructive plasticity.

The last chapter, "Quantum Derrida," takes up the work of Karen Barad on quantum physics, including quantum field theory, and shows how she develops an understanding of "hauntological materialism" that is indebted to Derrida. Here I explain how the phrase "every other is every other," or *tout autre est tout autre*, can be viewed in a quantum theoretical context, not simply an ethical or religious context. Each of the last three chapters contributes to a more materialist understanding of Derrida. In the Afterword to this book, "The Sins of the Fathers—A Love Letter," I reflect a little more personally on the issues of gender and patriarchy as they relate to philosophy, a topic that Derrida also addressed. Here I consider how Derrida came to serve as a kind of father figure for me, and how this is a symptom of the perpetuation of what Derrida calls "phallogocentrism." From a consideration of fathers and substitute father figures, including Caputo and my own teacher, Charles Winquist, I turn to more directly consider and reflect on the ideas of women philosophers such as Malabou, Julia Kristeva, Bracha Ettinger, Catherine Keller, and Katerina Kolozova. Here is another context in which Derrida's philosophy matters in ways that push us beyond writing and call for another paradigm, which for me is a paradigm of New Materialism.

To summarize, this book argues that we need to engage Derrida's later philosophy not from the standpoint of a turn, but from a new materialist perspective that treats politics and religion as material and spiritual practices. For Derrida's philosophy, there is a change, but it is not a simple change of theme or perspective. It is more a change of context. With Malabou, I argue that in the 1980s and 1990s we can see a shift from a motor scheme of writing to a motor scheme that she calls plasticity. If we want to be more faithful to Derrida's words, we could use a term such as machinic, automaticity, or teletechnology; or, following Arthur Bradley, we could call Derrida's post-writing motor scheme an "originary technicity."

Whether we want to call it technicity or plasticity, a careful consideration of Derrida's work attends to this shift that wrenches his philosophy out of an explicit context of writing. Writing is always already material, and Derrida was never a linguistic or transcendental idealist, but this new perspective opens up new ways to read Derrida's work afresh. In the late 1990s and early 2000s, religion and politics intersect and interact in a kind of political theology, even though Derrida wants to avoid this term because of its association with sovereignty in the work of Carl Schmitt.

In the twenty-first century, English-language studies of Continental philosophy have become more aware of and engaged with mathematics and the natural sciences with the work of Malabou, Alain Badiou, Bruno Latour, Quentin Meillassoux, François Laruelle, and the subsequent Speculative Realism and Object-Oriented Ontologies of Graham Harman, Ray Brassier, Levi Bryant, Timothy Morton, and others. I think this return to the natural sciences is important and necessary, but it risks caricaturing previous representatives of post-structuralism, such as Derrida, Deleuze, and Lacan. As I argue in Chapter 5, I think some of the newer realisms and OOOs go too far in their efforts to avoid subjectivism and subjectivity, and I think that they are more productive when read along with Deleuze, Lacan, and Derrida rather than against them.

My preferred theoretical perspective is New Materialism as opposed to Speculative Realism or OOO. New Materialism gives us tools to think about and better understand the complex relations among science, energy, money, power, politics, philosophy, and religion in the early twenty-first century. New Materialism is not a reductionist materialism, but a materialism based on body as theorized by feminism and cultural studies and energy as theorized by chaos and complexity sciences. Influenced by the philosophies of Alfred North Whitehead, Maurice Merleau-Ponty, and Gilles Deleuze, some of these theorists include Rosi Braidotti, William Connolly, Manuel DeLanda, Jane Bennett, and Isabelle Stengers.[3]

This New Materialism is sometimes described as a neo-vitalism, but if it is vitalist, it does not appeal to any transcendent immaterial or spiritual power, but stresses the vital immanence of material processes. In this book, I am suggesting that rereading Derrida in terms of New Materialism is fruitful, and this is the case in both Catherine Malabou's biological materialism and Karen Barad's work on quantum physics. Religion, politics, and ethics are complex material processes, and sovereignty, if there is such a thing, is not a unified entity but a distributed process that occurs along the edge of chaos in the form of a singularity. A singularity is not simply a determinate entity; it is a significant transformation or threshold where a new arrangement occurs as a result of a bifurcation or change.

With New Materialism, I am employing a theoretical perspective on Derrida's later philosophy that also can be characterized as a religious or spiritual materialism. Derrida does not endorse the term "materialism" because he views it in the older reductionist sense, but I think that there is a case to be made for reading Derrida as a new materialist. We could say that Derrida specifically avoids the word "materialism" for good and bad reasons. To the extent that one would think about a positive idea of materialism in Derrida's work, we might associate it with the notion of pure empiricism that he uses to think about a limit of and an outside to metaphysics, philosophy, and language. In "Violence and Metaphysics," Derrida claims that empiricism is "the *dream* of a purely *heterological* thought at its source. A *pure* thought of *pure* difference."[4] He says that we have to not renounce this dream of empiricism but to find another way to think about it.

What is crucial about this New Materialism for me is that it is a non-reductionist materialism; it employs ideas from the sciences and science studies, but it is not a form of scientism, which is something Derrida wants to avoid. It is above all not an atomic materialism. The New Materialism does not depend on any sort of dualistic relationship between what we call matter and what we call spirit. In the book I wrote with Jeffrey W. Robbins, *Religion, Politics and the Earth*, this spiritual materialism is based on energy transformation, where energy is at once fully material and fully spiritual.[5] Here we more explicitly draw on Deleuze, and Deleuze and Guattari's thinking of the Earth, in connection with Hegel's idea of Substance becoming Subject in the *Phenomenology of Spirit*. We suggest a transition from anthropocentrism to a thinking of and from the Earth. Earth here becomes thought of not just as substance, but precisely as subject, and we are at least in part an effect of the Earth. This is a Deleuzian interpretation of Hegel, where we posit a "Geology of Morals" that asks "Who Does the Earth Think it is?"

This new religious materialism, like most forms of New Materialism, is more explicitly influenced by Deleuze than by Derrida. But I am arguing that it is possible, and in fact important, to think about Derrida's philosophy from a more new materialist perspective. Rosi Braidotti, who coined the term "neo-materialism" in the 1990s, says that she wants to distinguish a materialist strand of post-structuralism from the more hegemonic linguistic strand of post-structuralism. "Thus 'neo-materialism' emerges as a method," Braidotti states, "a conceptual frame and a political stand, which refuses the linguistic paradigm, stressing instead the concrete yet complex materiality of bodies immersed in social relations of power."[6] This materialist strand is more informed by Georges Canguilhem, Michel Foucault, and Deleuze, whereas Derrida is usually seen as the major representative of the linguistic strand of post-structuralism.

From the perspective of the 1990s, especially in the English-speaking world, these two strands of post-structuralism indeed seemed oppositional and incompatible, but I think that they are less so now. Even if Derrida and Deleuze used very different languages, we have a much better sense of their complementarity. I don't think we need to rescue Derrida from this linguistic paradigm, but instead to see how the so-called linguistic paradigm was already a material paradigm, and that it functioned and functions far more broadly than is usually assumed. The linguistic turn seems excessive today to many scholars and philosophers, although we ignore the complexities of language at our peril, and we need to be careful not to lose sight of the many insights developed by poststructuralist thinkers not only about language but also about reality. Malabou's philosophy helps greatly in this effort to overcome the opposition between a linguistic and materialist poststructuralism. This is because she is more explicitly working in the tradition of Hegel, Heidegger, and Derrida, but she engages deconstruction with her new biological materialism in a way that crosses the two strands of poststructuralism that Braidotti identifies.

Both deconstruction and New Materialism go from human activities and interactions all the way down to the subatomic level. For Karen Barad, the self-touching of virtual particles that intra-act in quantum field theory is a queer example of *tout autre est tout autre*, and it expresses what she calls a hauntological materialism that is partly inspired by Derrida. Like Barad, I am envisioning Derrida as a theorist of deconstruction operating at a quantum level, not simply on large-scale actions such as human ethics and language. For the feminist theorist Vicki Kirby, who has both influenced and been influenced by Barad, "deconstruction discovers itself in enterprises such as cybernetics, biology, and chemistry."[7] In her book *Quantum Anthropologies*, Kirby asserts that "language is not a second-order representation or model of an absent world, but rather, an ontological energy through which the world makes itself known."[8]

Derrida is not a philosopher of materialism, but New Materialism is a viable perspective from which to engage Derrida. In an article on "Matter and Machine in Derrida's Account of Religion," philosopher of religion Michael Barnes Norton argues that attention to the machinic nature of religion in Derrida's work helps us think about the intrinsic materiality of religion and other processes.[9] According to Norton, religion is a material practice, and the machinic material practice of religion can be viewed in both positive and negative terms. Like Derrida, John D. Caputo is ambivalent about the term "materialism," although he does use it in *The Insistence of God*. Here Caputo argues that his positive idea of "cosmo-poetics could be formulated in terms of a 'religious materialism.'"[10] Here, religious materialism is about refusing—or deconstructing—the opposition between matter and spirit, form and force, body and brain.

In an essay on "Becoming Feces," radical feminist theologian Karen Bray cites Derrida's two columns from *Glas*, one on Hegel and one on Jean Genet. While the Hegelian column represents a desire for mastery and absolute knowledge, "Derrida's Genet column, the fecal column, runs alongside and betrays this search for mastery over 'unconscious' organic agencies." Bray imagines a new materialist, dishonored God "that becomes both brain and *shit*, vibrates in the material powers of decomposition and recomposition; and occupies with us and feels the black rage."[11] In this respect, she affirms a metabolic materialism of resistance that is informed and inspired by Derrida's work, among others, including feminism, queer theory, and affect theory.

In the case of Malabou's philosophy, her materialism concerns the notion of form, and I appreciate how she opens up and explains form in a complex and plastic manner. I do not subscribe to any simple opposition between form and force or energy. Malabou grounds force within form, whereas I see form more as a product of energy transformation. For me, New Materialism is fundamentally about energy. What is energy? Energy is a strange sort of object, and it stretches what we mean by object almost to the breaking point. Matter and energy are convertible at the square of the speed of light. In complexity theory, we see a phenomenon called self-emergence, where new objects emerge at thresholds of singularity sometimes called the edge of chaos. The Belgian scientist Ilya Prigogine calls certain kinds of dynamic systems that can be sustained far from equilibrium "dissipative structures."[12] These dissipative structures temporarily resist the overall tendency toward entropy so long as they are fueled by a continuous flow of energy.

Just as he doesn't write about materialism, Derrida also does not write explicitly about energy. What or where is energy for Derrida? I think there is a more direct connection between energy and force. For Derrida, force is a vital and necessary concept with which to think the limits of language. In his essay "Force and Signification," Derrida contrasts the title of his essay with a book on structuralism called *Form and Signification*. Here Derrida protests the "aesthetic which neutralizes duration and force" between two phenomena.[13] Later, in a brief reflection on Hegel's phenomenology, Derrida claims that "force is the other of language without which language would not be what it is."[14] I prefer the word "energy" to that of "force," but I think that we need to attend to the operation of the idea of force in Derrida's work. We could say that there is an underground current from Derrida's "Force and Signification" to his "Force of Law." Force is a kind of energy, a transformative energy, and Derrida mostly attends to the way that force disrupts language and dislocates meaning.

We could also, however, naively attempt to think force as energy more directly. Kirby's book helps us do just that, with her insistence that deconstruction is intrinsically tied to physical reality, and that language is an ontological energy. Carl Raschke also contributes to this perspective in his book *Force of God* when he says that "the Derridean undecidable amounts to an engagement with the *force* of the other."[15] For Raschke, the predominance of political theology means that every force is—provocatively—at least potentially a force of God. Energy is force, forces, and these forces make us—they are us. These energy forces are at one and the same time fully material and fully spiritual. Here is where materialism, religion, and politics, including the themes and concerns of political theology, intersect. And Derrida remains one of our most powerful and provocative resources to think about this intersection.

1

Reading Derrida Reading Religion

At his death in 2004, Jacques Derrida was the most famous philosopher in the world. Born in 1930 in El Biar, Algeria, Derrida was a Jewish child caught between Arab Muslim North Africa and European Christian France. Derrida never fully embraced his Jewish identity, calling himself ironically "the last of the Jews."[1] He moved to France to matriculate at the École Normale Supérieure, and then lived and taught in France and the United States, writing complex, influential and important texts. Although his early work treated religious ideas and themes, it was only in the 1990s that he began more explicitly discussing religion in positive terms. In the late 1990s and 2000s, readers in English began reflecting and commenting on the significance of this interaction, with the most famous interpreter being John D. Caputo, who published *The Prayers and Tears of Jacques Derrida* in 1997.[2]

In this opening chapter, I will survey some of Derrida's major writings, and draw out some of the primary religious themes. This survey will proceed chronologically, but it will not be exhaustive of Derrida's extraordinary and wide-ranging work, and it will provide context for later chapters that go into more depth on Derrida and others. Here I want to make two important claims. First, I think that Derrida was fascinated by religion during the course of his entire life, but he was never interested in traditional religious identity, dogma, or orthodoxy. As Edward Baring shows in his important study on *The Young Derrida and French Philosophy, 1945–1968*, "Derrida's thought can be understood within the context of French

Christian philosophy."[3] This is the case because of the importance of French Christian existentialists and their influence at the École Normale Supérieure, and the division there between the Catholic Christians and the Communists aligned with Louis Althusser.[4] Derrida was sympathetic to both sides, but he aligned himself with neither, choosing to work on Husserl as a philosopher who was compatible with both groups.[5]

Derrida's interest in religion is always balanced by a critical suspicion about the ways in which religion has been used to oppress people and obscure knowledge. Derrida's philosophy maintains this delicate balance, thinking along the knife-edge between faith and doubt. His religion is a "religion without religion," as Caputo puts it, because he wants to think about religion in itself as a pure possibility beyond any determinate or phenomenal form of religion. Derrida comes to name this first type of religion a kind of messianicity, but this messianicity cannot be equated with the presence of a Messiah. This messianicity is Jewish in a way, and a lot of work has discussed Derrida's complicated relationship to Judaism, but it is not a simple Judaism partly because Derrida does not think that one can avoid thinking about Christianity if one is thinking about religion.[6] That doesn't mean one has to be or become Christian, but Christianity is always there to influence what we mean by religion, for good and for bad. Derrida wants to deconstruct or dislocate substantial religious identities, but he affirms a certain kind of faith that he believes is irreducible or undeconstructible.

Derrida does not believe that you can simply have a generic, indeterminate religion without a specific, historical religion; he affirms that one never exists without the other. At the same time, he wants to think what religion means in excess or outside of this or that particular religious tradition, and that is what deconstruction is about. Deconstruction attends to that which exceeds any particular tradition or structure, and shows how that excess is both included within the tradition and excluded from it in its very formation. For Derrida, the pure possibility of religion concerns the promise, the possibility of making a promise and being responsible to and for another person. This responsibility is ethical in many respects, and Derrida is profoundly influenced by the Jewish philosopher Emmanuel Levinas, but at the same time ethics is always at the same time religious and political, because these boundaries are not fixed but permeable.

The second point I want to emphasize here concerns the idea of a "turn" in Derrida's philosophy from early to late. This suggestion of a turn corresponds to the distinction between early and later Heidegger, as a shift from *Dasein* as the being who asks the question of being, to the notion of being in itself that shows itself in an appropriating event, or *Ereignis*. I think that in the case of Derrida, this language of a turn is overstated. Yes, there is a definite shift in emphasis in Derrida's philosophy after 1989, but this is already prefigured in his work in the 1970s, and the shift in emphasis does not imply any sort of repudiation of Derrida's earlier work in the 1960s and 1970s. Derrida has always been interested in religion, but he comes to write about it more explicitly in relation to ethics and politics, and the co-incidence of these three themes after 1989 becomes more prominent in his later work. 1989 is the year of the presentation of his important essay "Force of Law: The 'Mystical Foundation of Authority,'" at a conference at Cardozo Law School in New York City. If there is a turn in Derrida's philosophy toward an explicit engagement with religion, it can be traced to this essay. The later works dealing with religion all stem from this important paper, which I will discuss shortly.

If there is a turn in Derrida's philosophy, however, there is not just one. At a conference devoted to his work in 1982, Derrida himself reflects on a shift in his own thought, from "guarding the question," "insisting on the priority of an unanswerable question," which is *différance*, to responding to a call to support the other, as "that which must be differed-deferred so that we can posit ourselves, as it were."[7] This other is both wholly other and also at the same time a particular, historical concrete other. As Derrida expresses it in *The Gift of Death*, "*tout autre est tout autre*," that is, every other (one) is every (bit) other. Even if there is this shift, though, it is not a repudiation of his earliest work on Husserl and grammatology because this early work has always been engaged with liberating the otherness of phenomenology, grammatology, and language. There is no question without the other, and no other without the posing of a profound question that implicates me. In his book *Derrida and Theology*, Steven Shakespeare claims that "Derrida's thought invites the coming of the other, the address of the other, and this is an irreducibly religious motif."[8] Insofar as Derrida's philosophy constantly and consistently treats the relation with the other, it is marked by a religious sensibility. Here Derrida follows the work of his close friend and collaborator, Levinas, whose philosophy is also primarily concerned with the other, and is also entwined with religion.

So I resist the language of a strong turn in Derrida's philosophy, especially a turn to religion. At the same time, there are important emphases in his later works to which I will attend. And in terms of the book as a whole, I suggest that Derrida's later emphasis on religion, ethics, and politics can be read not as a turn or reversal but as an opening up of deconstruction beyond writing in a narrow, technical sense. As Carl Rashke claims in his impressive book *Force of God*, "this opening amounts less to a 'turn' away from pure deconstruction to matters ethico-political and religious than a kind of epochal elucidation of what has been tacit but not apparent in his philosophical enterprise all along."[9]

Many critics of Derrida associate his philosophy with the linguistic paradigm that came to dominate structuralism and post-structuralism, and read *différance* mainly as a linguistic operation. Language and writing are what Catherine Malabou calls the motor scheme of mid—to late–twentieth-century philosophy, and this scheme has receded in significance over the past couple of decades. My argument is that there is still reason to read Derrida beyond and even after the paradigm of "writing," at least in a technical sense. The arguments about Derrida and religion, especially his affirmation of this or that religion, are less important than understanding how his work demonstrates that deconstruction begins as a form of writing but then opens up beyond any narrow definition of writing.

This chapter highlights some general themes of Derrida and religion that carry through his life and work. First I consider some of the religious resonances that animate his earlier texts. Derrida cut his teeth on the philosophy of Edmund Husserl, and his first publication in 1962 was a translation of Husserl's *Origin of Geometry*, along with a long, critical introduction. In this work, Derrida develops his project of pushing philosophy through and beyond phenomenology, and this effort culminates in his longer work on Husserl, *Voice and Phenomenon*, published in French in 1967. Derrida wants to show the limits of phenomenology by noting its dependence on speech as a kind of living present, on which phenomenology bases its understanding of the Idea. An Idea is both historical and ahistorical at the same time, which means that it is divided from itself and cannot be a pure, consistent idea. Husserl's Idea of God is equivalent to the Idea of Logos, and this Logos or Speech is at the same time constituted in and through history and provides history with its telos and transcendental meaning.

Derrida writes:

> God speaks and *passes through* constituted history, he is *beyond* in relation to constituted history and all the constituted moments of transcendental life. But he is *only* the Pole *for itself* of *constituting* historicity and *constituting* historical transcendental subjectivity.[10]

This is a complex argument, but Derrida is showing how an understanding of God animates Husserl's phenomenology in a problematic way because it is aligned with traditional understandings of God as Logos. Here God is equivalent to the Platonic Idea, which both participates *in* history and at the same time transcends history in an idealist way. Derrida questions the consistency of this articulation and suggests that even in his later work Husserl cannot escape the paradoxical formulation of an idea in transcendental, and even implicitly theological, terms. Derrida's early work is seen as more hostile to religion because he is showing how the fundamental opposition between speech, as closer to the ahistorical Idea, and writing, as the fall of the Logos into history, deconstructs, and how writing always already inhabits our notions of speech.

In *Of Grammatology*, published in French in 1967, Derrida further develops his analysis of the interrelation between speech and writing by way of a reading of Rousseau's *Essay on the Origin of Languages*. Western philosophy is religious insofar as it desires pure self-presence in the form of speech or Logos, and it wants to separate and segregate writing as a form of corruption, decay and death. Writing is an external supplement that delays the self-presence of speech, that must be kept exterior to living speech. Derrida demonstrates that what we call writing already inhabits and animates speech. In *Of Grammatology*, he develops the conception of "archewriting" as a name for that which makes speech and writing possible, and prevents them from ever fully closing in on themselves.

Writing has to do with spacing, deferral, and delay. It is what prevents full self-presence, and it disrupts, deconstructs, and opens up metaphysics to something other than itself. In the middle of the twentieth century, the sciences of anthropology, psychoanalysis, and linguistics demonstrate the profound importance of the written signifier, and this renders philosophical logocentrism problematic, as Derrida shows in his work on Western philosophy from Plato to Heidegger. Derrida says that the "notion of the sign always implies within itself the distinction between signifier and signified."[11] The problem, however, is that "fundamentally nothing escapes the movement of the signifier and that, in the last instance, the difference between signified and signifier is *nothing*."[12] If there is no absolute difference between signifier (writing), and signified (concept, speech as a whole), then there is no transcendental signified.

The idea of a transcendental signified is here associated with God, or with what Heidegger calls "ontotheology," where the idea of God governs and supports the totality of beings in a metaphysical way. Readers of Derrida who maintain a traditional conception of God as supreme being read this claim that "there is no transcendental signified" as declaring that there is no God. This conclusion is an interpretive leap, but Derrida does break with traditional concepts of God, and he shows how our linguistics and semiotics often invoke this logic of God as a transcendental signified, even if we are atheists.

If the difference between signifier and signified is not absolute, then Derrida claims that even categories like Heidegger's ontological difference "are not absolutely originary. Differance by itself would be more 'originary,' but one would no longer be able to call it 'origin' or 'ground,' those notions belonging essentially to the history of ontotheology, to the system functioning as the effacing of difference."[13] Here is Derrida's famous neologism, differance, in French *différance*, which is simply the word difference spelled with an *a* instead of an *e*.[14] In his famous essay "Différance," included in *Voice and Phenomenon* as well as the later book *Margins of Philosophy*, Derrida claims that to differ is both to be different in a static sense and also to defer in a temporal sense. By distinguishing his word differance with a mark that can be read but not heard in oral speech, he is designating a "temporization" at work in difference. Differance is temporalization and spacing, and therefore it is "the becoming-time of space and the becoming-space of time."[15]

Differance is what disrupts self-presence, and it testifies to a trace of signification that leaves a mark that makes signification possible and refuses to allow it to come to any final completion. Language, history, philosophy, meaning and sense: These are all ongoing transforming and transformative processes that make us even as we use them to get a handle on what we are doing with them. Differance is not a simple origin because it means that there is no simple unified origin. Differance precedes God and makes God as Logos impossible in any pure or undivided way. This is why Derrida is seen as hostile to religion in his early works.

If Derrida's early work seems to dismiss traditional ideas of religion, the essay that introduced him as a rising star to an American audience expresses the important notion of an event, which later becomes infused with religious meaning. In 1966, Johns Hopkins University held a conference on "The Languages of Criticism and the Sciences of Man," featuring well-known theorists such as Roland Barthes, Paul de Man, Jean Hyppolite, and Jacques Lacan. This conference was intended to help introduce and clarify structuralism for intellectuals in the United States. In some ways, Derrida's presentation, "Structure, Sign and Play in the Discourse of the Human Sciences" upstaged his elders and signaled a transition toward what became known as post-structuralism.

The first sentence of Derrida's talk, which was given in French, reads: "Perhaps something has occurred in the history and concept of structure that could be called an 'event.'"[16] In the rest of his famous essay, Derrida develops some of the implications of this idea of an event. An event distorts and disrupts structure, even as it serves to make structures possible. An event is not simply an event contained within a structure, but more fundamentally an event that founds a structure. An event is not a simple origin, but a working of differance in and through events and structures. A structural field cannot be totalized, and this is because "this field is in effect that of play, that is to say, a field of infinite substitutions only because it is finite." There is no center or transcendental signified "which arrests or grounds the play of substitutions."[17] Even if there is no God who functions to arrest or ground the play of substitutions, the event offers a kind of religious interruption of systemic structures that animates and inspires post-structuralism. In his recent work, John D. Caputo takes up the idea of an event as a theological occurrence that manifests the insistence of God rather than the existence of God, as discussed in Chapter 6.[18]

Before turning to Derrida's later work, to see where and how religion becomes more central to it, I will look briefly at another early essay by Derrida. His important engagement with Levinas, "Violence and Metaphysics," was originally published in 1964 and is included in the collection *Writing and Difference*. This critical treatment of Levinas touches on religious themes, and later leads to associations and questions about the relation between deconstruction and negative theology. "Violence and Metaphysics" is an extraordinary encounter with Levinas, who in the early 1960s was not very well known even in France, and Derrida is incredibly affirmative of Levinas's philosophy even as he criticizes it at certain points.

Derrida begins by posing the question of the death of philosophy, and then moves to take on the challenge of Levinas's thought as both a challenge to philosophy as well as a renewal of it. He claims that "this thought calls upon the ethical relationship—a nonviolent relationship to the infinite as infinitely other, to the Other—as the only one capable of opening the space of transcendence and of liberating metaphysics."[19] Derrida questions whether Levinas can fully liberate metaphysics, but he does affirm the value of making ethics and metaphysics "flow into other streams at their source," namely Jewish streams.[20] Levinas is deeply engaged with philosophy, and he is opposing philosophy in the name of ethics and religion. Derrida states that "the ethical relation is a religious relation. Not *a* religion, but *the* religion, the religiosity of the religious."[21] Religion names the ethical relation between the human and God, or the human and the face of the Other, which is a relation of asymmetry and radical alterity.

Derrida questions the ability of Levinas to free his philosophy from Heideggerian Being as well as Derrida's own deconstruction of the binary of speech and writing, but he does argue that Levinas offers a thinking of divinity within the context of language and history rather than one that is solely outside in the manner of negative theology. Derrida says that "negative theology never undertook a Discourse with God in the face to face, and breath to breath, of two free speeches," in the way that Levinas does.[22] Philosophy is caught "between original tragedy and messianic triumph," and the tension of this questioning animates Levinas's philosophy even as he struggles to free himself from the Greek, tragic pole.[23] Levinas appeals to a language of God that would found or sustain a pure empiricism of the pure thought of pure difference, but he has to pass through language to do this. In some ways, Levinas tries to do what negative theology does, but Derrida judges Levinas's attempt in more positive terms. Levinas must borrow Greek logic to flesh out Jewish experience, and this is not an impurification but a necessity, a necessity that makes history, and makes us subjects of this history where we no longer know whether we are simply Greeks or Jews.

Despite Derrida's clear demarcation of the distance of his philosophy from that of negative theology, in the 1970s and 1980s many readers suspected some sort of connection. In an address given in Jerusalem in 1986, "How to Avoid Speaking: Denials," Derrida argues that deconstruction is not a type of negative theology because negative theologies generally appeal to some sort of hyper-being beyond being. Derrida's essay is partly a response to and an implicit critique of Jean-Luc Marion's book *God Without Being*. Derrida suggests that Marion appeals to a God beyond or above Being in a way that accords with classical negative theology such as that of Pseudo-Dionysus. Despite the desire to move beyond being, there remains an "analogical continuity in the rhetoric, grammar, and logic of all the discourses on the Good and on what is beyond Being."[24] Although Derrida refuses any conventional theological identification of deconstruction, he also suggests provocatively that deconstruction in the form of *khora* is at work within these more traditional theological discourses, including that of negative theology. *Khora* is a Platonic term for a material receptacle, a matrix or place, and Derrida suggests that this logic of *khôra* occurs in prayer, both apophatic and kataphatic prayer.[25] *Khôra* is the locus of prayer that is addressed to an other or Other. This denial of negative theology is not a simple denial, but in French a *de-negation*, which dislocates and displaces the denial-affirmation structure in the name of *khôra*.

During the later 1980s, two controversial intellectual events occurred that I think helped encourage Derrida to be clearer about his ethical and political commitments. One was the continuing controversy over Heidegger's association with Nazism in the 1930s. Derrida was heavily influenced by Heidegger's philosophy, but he was always critical of Heidegger's politics. Much of the historical work on Heidegger's affiliation with the Nazi party had already been established, but a book by Victor Farías on *Heidegger and Nazism* was published in French in 1987 and created a new storm of accusations.[26] Also in 1987, a scandal erupted as a result of the posthumous discovery of a number of writings by Paul de Man for a collaborationist Belgian newspaper in the 1940s, some of which were explicitly antisemitic. Derrida was a close friend of de Man, and he was stung by the fury of the scandal and the way that it was used to demonize postmodern literary criticism. I speculate that these controversies surrounding Heidegger and de Man provide part of the context for the writing of "Force of Law."

In 1989, Derrida was invited to give an address at the Cardozo Law School by Drucilla Cornell, on the topic of "Deconstruction and the Possibility of Justice." His essay, "Force of Law: The Mystical Foundation of Authority," distinguishes justice from law, and affirms that "*deconstruction is justice.*"[27] Law is something determinate, and law is always instituted in the name of justice, but it can never exhaust justice. Justice is not a determinate thing; it is not deconstructible because it does not exist as such. According to Derrida, "deconstruction takes place in the interval that separates the undeconstructibility of justice from the deconstructibility of law. Deconstruction is possible as an experience of the impossible, there where, even if it does not exist, if it is not *present*, not yet or never, *there is* justice."[28] Here Derrida aligns deconstruction with the experience, promise, or invocation of justice, which takes place beyond the literal existence of the law. There is never law without justice, just as there can never be justice without any law, but they cannot be collapsed into each other.

The affirmation of deconstruction as justice and of justice as undeconstructible is an ethical and a political affirmation. The experience of justice is an aporia, a paradox or impassable passage that has no simple solution. This shift of emphasis in Derrida's philosophy opens up a new period of work in which he writes more explicitly and affirmatively about ethics and politics, which cannot be completely separated from one another. At the same time, this affirmation of the ethical and political aspects of deconstruction is tied to an acknowledgement of the *religious* nature of deconstruction.

Why is this the case? For Derrida, it has to do with the subtitle of the essay, "The Mystical Foundation of Authority," which is a phrase from Montaigne that is related to an essay by Walter Benjamin. In "Force of Law," Derrida is writing about a famous essay of Benjamin's, called "Critique of Violence." The German word for violence, however, is *Gewalt*, which also means force. In his essay, Benjamin distinguishes between different kinds of force or violence: lawmaking violence and law-preserving violence. Benjamin calls lawmaking violence a kind of mythic violence because it founds a new order or way of living. Most determinate laws are law-preserving insofar as they protect and preserve already established frameworks. At the same time, Benjamin talks about a kind of divine force that would be more destructive, and exist outside the law and takes place as "revolutionary violence."[29]

I think Derrida sees deconstruction as a kind of revolutionary force, but he does not separate it from lawmaking, mythic force to the extent that Benjamin does. The point is that in any new lawmaking, or any institution of a new authority, there is a "mythic" positive violence and a "divine" negative violence. Derrida argues that we should identify the divine force of lawmaking institution as justice, and see deconstruction as a non-violent force that accompanies the violence of any lawmaking institution. Any new foundation of law occurs in the name of justice, and justice always has a mystical element to it.

Religion, ethics, and politics are intertwined in a complex web of practices and significations. The distinction between determinate-deconstructible law and aporetic-undeconstructible justice is repeated in religious terms in Derrida's formulations of "messianicity without messianism" in *Specters of Marx* and "religion without religion" in *The Gift of Death*, culminating in the lecture on "Faith and Knowledge." In *The Gift of Death*, published in French in 1992, Derrida reflects on the religious nature of contemporary philosophical discourse. He considers Kierkegaard's *Fear and Trembling* as a retelling of a story that concerns a gift of death from Abraham to Isaac and from God to Abraham. In his interpretation of *Fear and Trembling*, Derrida generalizes Abraham's situation in response to God's call to sacrifice Isaac. He writes:

> As soon as I enter into a relation with the other, with the gaze, look, request, love, command, or call of the other, I know that I can respond only by sacrificing ethics, that is, by sacrificing whatever obliges me also to respond, in the same way, in the same instant, to all the others. I offer a gift of death, I betray, I don't need to raise my knife over my son on Mount Moriah for that.[30]

Our ethical obligations to each other require that we sacrifice someone else to fulfill them, which is the paradox and ruin of conventional ethics. Kierkegaard posits a religious category beyond ethics, where absolute relation to the single individual holds, but Derrida wants to universalize Kierkegaard's idea of the religious.

Derrida says that "*tout autre est tout autre*," which means that "every (other) one is every (bit) other."[31] He explains that "if every human is wholly other, if everyone else, or every other one, is every bit other, then one can no longer distinguish between a claimed generality of ethics that would need to be sacrificed in sacrifice, and the faith that turns towards God alone, as wholly other, turning away from human duty."[32] This means that any absolute opposition between God as wholly other and the human being as a distinct other person deconstructs, and so does any ultimate separation of religion and ethics. I will return to this powerful formulation and draw out more of its implications in later chapters.

Derrida argues that neither Kierkegaard nor Levinas can maintain a consistent distinction between ethics and religion. "In the two cases the border between the ethical and the religious becomes more than problematic, as do all attendant discourses," he writes. And "this applies all the more to political or legal matters."[33] The point is that the religious, the ethical, and the political are imbricated upon each other in Derrida's writings in the 1990s. Furthermore, in each case Derrida wants to think the possibility of a religion without religion, a pure promise, and a democracy to come, in a way that subtracts from the determinate instantiations of these discourses. At the same time, we can never free ourselves from the determinations of these political, ethical and religious histories, laws and effects.

In *Specters of Marx*, published in French in 1993, Derrida provides a re-reading of Karl Marx and shows how his thought still haunts our world despite the collapse of socialism and the apparent triumph of global capitalism. He points out the gap between liberal democracy and the ideal of democracy, and argues that this gap is irreducible because democracy means for Derrida a democracy *to come*, which is not a future present but an infinite promise of justice. Democracy to come names "the opening of this gap between an infinite promise . . . and the determined, necessary, but necessarily inadequate forms of what has to be measured against this promise."[34]

There is a quasi-mystical or religious nature of this gap and this promise, which Derrida calls spectral, and he connects this to the "strange concept of messianism without content, of the messianic without messianism, that guides us here like the blind."[35] Messianism possesses too much content, so it is determinate and deconstructible, and what's worse, it collapses the gap between the infinite promise of justice, democracy, or religion, and its actual present form. Derrida invokes a stripped-down, deserted messianicity as a perspective through which to read Marx's eschatology, which does not work as a prediction of a future communist present. Still, Marx's work continues to haunt us in its messianic spectrality, calling us to a different just future.

This work on Levinas, Kierkegaard, and Marx in the early 1990s lies in the background of Derrida's famous Capri lecture of 1994, published as "Faith and Knowledge: The Two Sources of 'Religion' at the Limits of Reason Alone." Derrida wants to think about the pure possibility of religion apart from any and every determinate religion, while recognizing the priority of Christianity in producing the idea and concept of religion in general that we consider. Religion plays a role in what Derrida calls "globalatinization," which is a mash-up of globalization and latinization, or what we call the *West*.

According to Derrida, religion names "the convergence of two experiences that are generally held to be equally religious:

1 the experience of *belief*, on the one hand . . . ; and
2 the experience of the unscathed, of *sacredness* or of *holiness*, on the other."[36]

These two sources make up what we call religion, and the division into two sources already complicates the idea of the unscathed, which implies a certain unity or lack of division. Derrida is suspicious of this second source, and he suggests that our desire for the sacred is significant but not ultimately fulfilled.

When we desire the sacred as the unscathed, we resist anything that appears to threaten this experience, and this sets up a situation that Derrida calls auto-immunity, because a social body reacts to reject what it believes threatens it from outside, but this very reaction is actually more dangerous and destructive to the body than the so-called disease. This is just like an auto-immune disease that destroys an organic body when the immune system gets out of control. So there is a "space where all self-protection of the unscathed, of the safe and sound, of the sacred (*heilig*, holy) must protect itself against its own protection, its own power of rejection, in short against its own, which is to say, against its own immunity."[37] Here reactionary religious fundamentalism is aligned with science and technology, or what Derrida calls tele-technoscience, because these conservative forms of religion both oppose modern science and reason, and at the same time, make use of the tools of science and technology to oppose modern reason and to advance their own sense of the sacred as unscathed.

Derrida is critical of the second source of religion, the experience of the unscathed or the holy. He is less critical of the first source, the experience of belief, even though he wants to be extremely careful about belief as well. There is something about belief as trust that accompanies every relationship and every encounter; it is irreducible and ineliminable. Derrida is suspicious of simple, straightforward belief, but when belief is crossed with a complex understanding of the unscathed (the understanding that the sacred is never purely unscathed or undivided) it can produce a kind of faith that Derrida affirms.

There is something religious about every conception of community, and that is both negative and positive for Derrida. Religious faith is composed of the two sources, belief and the unscathed, and cannot be reduced simply to either one. Religion is ambiguous and ambivalent: On the one hand, religion as auto-immunity threatens to shut down and close off community, but on the other hand, religion as the bond of belief, or in Kantian terms dignity, is an absolute value of life "above and beyond the living, whose life has absolute value by being worth more than life."[38]

The exposure of life to what is more than life is not a simple belief, but a complex and contested experience of faith that tests and contests itself. According to Derrida, "this self-contesting attestation keeps the auto-immune community alive, which is to say, open to something other and more than itself."[39] Religion also indicates the convergence of these two sources in what Derrida calls witnessing, which is at stake in "every address to the other."[40] Ultimately I do not think Derrida sees the two sources as equal, preferring the experience of belief as fiduciary trust, while at the same time complicating this trust by recognizing that we can only trust something or someone if it is possible for this trust to break down or not be reciprocated.

In his book *Miracle and Machine*, Michael Naas examines Derrida's entire philosophy through the lens of this essay "Faith and Knowledge." Naas quotes Derrida as saying that "there is no social bond without faith," but then explains that for Derrida faith is not just a positive determinate relation but fundamentally a disruption or interruption.[41] Derrida claims that he believes in faith, but at the same time "Derrida is quite clear about the disruptive nature of this faith."[42] Derrida's faith contests itself, and this self-contestation is the essence of faith, at least responsible faith.

Jacques Derrida is one of the most important philosophers of the twentieth century, and the thinking of religion occupies a significant portion of his work. Derrida's understanding of religion is complex, ambiguous, and ambivalent. Religion is always contested, and it is bound up with the ethics of our responsibility to the other, as well as the politics of our living together with each other. Derrida does not endorse a particular form of religion, but he also does not renounce the religious as an integral part of what it means to be human.

Beyond the simple questions of whether Derrida is or is not religious, or what his relationship to Judaism, Christianity, or Islam is, in this chapter I have tried to sketch out an overview of Derrida's understanding of religion in more general terms. My claim is that religion, not only religion but certainly also and significantly religion, becomes a mode for Derrida to articulate deconstruction beyond the constraints of writing in a narrow sense. It's not that Derrida turns against writing, but in his later work he writes in ways that allow deconstruction to function in broader contexts. Of course, writing for Derrida originally has a much broader context, but that fact is sometimes lost on readers of Derrida in English.

In the next chapter, I will focus more specifically on Derrida's critical understanding of Christianity. He does not endorse Christianity in any straightforward way, but he also sees how difficult it is to simply oppose Christianity. He distances himself from his friend Jean-Luc Nancy's appropriation of his work as indicating "the deconstruction of Christianity." Derrida hesitates to affirm this deconstruction of Christianity as the destiny of deconstruction, although his goal is not to overcome Christianity, which is impossible, but to *survive* it.

2

Surviving Christianity

To what extent is deconstruction essentially a deconstruction of Christianity, as Jean-Luc Nancy suggests in his two-volume project on "The Deconstruction of Christianity": *Dis-Enclosure* and *Adoration*? In a related sense, what is the connection between poststructuralism and postsecularism? Although Derrida endorses and embraces the term "deconstruction," he keeps his distance from the word "postmodern." At the same time, there are resonances of a kind of reflection on postmodernism in *The Post-Card*, where the prefix post refers to a kind of "beyond" in the sense of Freud's *Beyond the Pleasure Principle*. Derrida writes that "to bind, therefore, is also *to detach*, to detach a representative, to send it on a mission, to liberate a missive in order to fulfill, at the destination, the destiny of what it represents. A *post* effect."[1] This post effect refers to a detachment that is not beyond in any spatial sense or after in a linear temporal sense, but is in some uncanny way already at work in the process to which it refers. Postmodernity is a beyond (detachment) in and of (the binding that is) modernity. Poststructuralism is a beyond in and of structuralism.

In a related sense, we could say that postsecularism is a beyond in and of secularism. In his later work, Derrida explicitly engages with religious themes, prompting many readers to claim that there is some sort of a "turn" post-1989 that rivals Heidegger's turn. I think that this language of a turn is overwrought, as suggested in the previous chapter, but I also think some of Derrida's protests to the contrary, where he cites himself at length treating similar themes from the beginning of his career, are also a little overdone. There is a shift in emphasis that emerges as what Catherine Malabou calls writing as a motor-scheme recedes, and Derrida's engagements with religious, ethical, and political themes appear more direct, abstract and ungrounded.

A different, earlier account of a shift in Derrida's philosophy is related by Gayatri Chakravorty Spivak, the first translator of *Of Grammatology* into English in 1976. In *A Critique of Postcolonial Reason*, Spivak says that in 1982 at a conference on Derrida's works at Cerisy-la-Salle, he "described a moment in his own work" that "was a turn from 'guarding the question'— insisting on the priority of an unanswerable question, the question of *différance*—to a 'call to the wholly other'—that which must be differed-deferred so that we can posit ourselves, as it were."[2] There is a shift from the priority of keeping the question alive to that of supporting the other, answering the call of responsibility. This turn occurs earlier than the important 1989 essay "Force of Law" and the works that take up religious themes more directly in the 1990s, and it suggests that we should attend to multiple shifts within Derrida's trajectory even as we look for lines of continuity.

I imagine that this turn to the other in the 1970s is also connected to Derrida's statement near the beginning of his presentation at the conference on Religion at Capri in 1994 that was expanded and published as "Faith and Knowledge: The Two Sources of 'Religion' at the Limits of Reason Alone." Derrida remarks: "No Muslim is among us, alas, even for this preliminary discussion, just at the moment when it is towards Islam, perhaps, that we ought to begin turning our attention. No representative of other cults either. Not a single woman!"[3] Here his exclamation about the lack of women can be connected to his work on the question of phallogocentrism and its exclusion of women in the 1970s, including his analysis of Nietzsche's apparent misogyny in *Spurs*.[4]

Insofar as there exists a turn, or better an inflection point that can be traced to 1989, it involves the working out of a motor scheme of writing and beyond it, and then into a newer motor scheme that Malabou calls plasticity. This contrast will be developed further in Chapter 7, in terms of a more substantial engagement with Malabou's interpretation of Derrida. The transition from writing to plasticity as an underlying motor scheme is less a conscious turn and more of an underlying context. Furthermore, it is not exclusively connected to religion, although it is also not entirely detached from it, insofar as religion is a constant theme of Derrida's work, but most explicitly and intensively so in the 1990s, with works such as *The Gift of Death*, *Spectres of Marx*, and "Faith and Knowledge." Furthermore, at the invitation of John D. Caputo, Derrida participated in three conferences—in 1997, 1999, and 2001—on Religion and Postmodernism held at Villanova University. Again, I want to contextualize this focus on religion, and the ways in which religion figures more and more explicitly in Derrida's work, as a way to think about deconstruction after writing as a specific paradigm or motor scheme has receded.

So for the later Derrida, there is religion, a word that is "the clearest and most obscure," and it has two sources, "the convergence of two experiences," one of which is the fiduciary experience of belief, faith or credit, and the other is "the experience of the unscathed, of *sacredness* or of *holiness*."[5] Another way to think about this convergence is to acknowledge that we can never simply avoid or get beyond religion, even if there is always this beyond (a detachment over and above binding) in and of religion. Any simple opposition between the religious and the secular deconstructs. This insight problematizes the dialectical-historical interpretation of modernity as the triumph of secularism over religion and then the view that the return of religion in philosophy, culture, and politics somehow instantiates a postsecular viewpoint that dispenses with the modern secular. Postsecularism is a corollary to what is often called postmodernism: Modernism cannot maintain its separation and elevation from what is denigrated as not-modern, and secularism is unable to rigorously enforce a boundary between the religious practice of a private faith and the public expression of secular reason and law.

Derrida advances a thinking of deconstruction that he affirms as a form of justice because it renders unstable all determinate expressions of law even as it makes them possible. Our predominant common (European) culture, Derrida acknowledges, is "manifestly Christian, barely even Judaeo-Christian."[6] If our common culture is Christian, and Christianity as the universal form of religion dominates our "globalatinization," then wouldn't the primary task in a world ridden with religious conflict be the deconstruction of Christianity?

This becomes Jean-Luc Nancy's project, to demonstrate how and why deconstruction involves the deconstruction of Christianity. In his 1995 essay "The Deconstruction of Christianity," later republished in *Dis-Enclosure*, Jean-Luc Nancy identifies Christianity with the heart of the West. He declares that "Christianity is inseparable from the West." Christianity is coextensive with the West as West insofar as both refer to "a certain process of Westernization consisting in a form of self-resorption or self-surpassing."[7] Nancy cites Marcel Gauchet's book *The Disenchantment of the World*, in which Gauchet argues that Christianity is the religion that leads religion beyond religion.

Both Christianity and the West, which are essentially the same phenomenon, consist in a self-carrying to the limit, and then giving themselves up in order to be true to "the depths of our tradition."[8] This movement is a kind of deconstruction, although Nancy also associates it with Hegelian *Aufhebung*—"letting go of the West and letting go of Christianity."[9] So it is the kenotic self-emptying or progressive self-overcoming of Christianity that issues in Western culture and thought, "a heart that risks being, if I dare say so, Christian."[10]

What is deconstruction, according to Nancy? He claims that deconstruction "is shot through with Christianity," and it is only possible from within a Christian horizon. "To deconstruct," Nancy says, invoking Heidegger's term *Destruktion*, "means to take apart, to disassemble, to loosen the assembled structure in order to give some play to the possibility from which it emerged but which, qua assembled structure, it hides."[11] Despite Heidegger's distance from Christianity, Nancy concludes that deconstruction is a Christian operation. Deconstruction "is Christian because Christianity is, originally, deconstructive, because it relates immediately to its own origin as to a slack [*jeu*], an interval, some play, an opening in origin."[12]

This logic of Christian exceptionalism loosely follows not only Gauchet, but also Réné Girard's claim that Christianity is the religion that exposes the workings of scapegoating and mimetic violence that drive most religions and cultures. For Girard, in his groundbreaking book *Violence and the Sacred*, human violence is caused by sacrificial desire that seizes on a scapegoat to sacrifice for the sake of the social body. Later Girard argues that Christianity takes this process to an extreme and exposes sacrificial violence by having Christ take on all sins and becoming the scapegoat for humanity.[13] Christianity exposes and overcomes sacrificial violence for Girard, just as Christianity overcomes and leads beyond religion for Gauchet. In a similar and implicit way, the West overcomes ordinary non-western culture and surpasses it in the theories of Girard, Gauchet, and Nancy. We participate in this self-surpassing by acknowledging, deconstructing, and letting go of it. Part of this acknowledgment would be the insistence that deconstruction, like the West, is essentially Christian, and is impossible without Christianity. Christianity is the exemplary, self-surpassing religion, and it is linked with the dominant world culture.

In this essay, which was originally presented in 1995, a year after the famous seminar at Capri, Nancy claims that we must think Christianity today, "that the Christian or Christianity is the *thing itself* that is to be thought" rather than avoided.[14] Nancy affirms here the possibility of an atheism "that contemplates the reality of its Christian origins."[15] Although Derrida also acknowledges the centrality of Christianity to globalatinization or Westernization in "Faith and Knowledge," he is much more ambivalent about this process than Nancy, and does not go so far as to affirm it unreservedly, especially the teleology of Christianity's surpassing and self-surpassing nature.

In his 2000 book on Nancy, *On Touching—Jean-Luc Nancy*, Derrida reflects on Nancy's extraordinary book *Corpus*, and on the issue of a deconstruction of Christianity more generally. Derrida acknowledges the Christian origin of the concept of deconstruction, from Luther's *deconstructio* to Heidegger's *Destruktion*. At the same time, Derrida distances himself from Nancy's project, warning that "'The Deconstruction of Christianity' will no doubt be the test of a dechristianization of the world—no doubt as necessary and fatal as it is impossible . . . . Dechristianization will be a Christian victory."[16] Neither Derrida nor Nancy believe in Christianity in any traditional manner, and they both acknowledge the complicity of deconstruction with Christianity, but whereas Nancy wants to affirm this in a more straightforward dialectical way, Derrida wants to undo it at the same time.

Part of the problem is the hyperbolic nature of Nancy's task. Derrida claims that "a certain Christianity will always take charge of the most exacting, the most *exact*, and the most eschatological hyperbole of deconstruction, the overbid of 'Hoc est enim corpus meum.'"[17] Nancy's hyperbolic project can always be outbid by a counter-Christian hyperbole that would subsume deconstruction into Christianity in a more conventional sense. Derrida advises Nancy that "a 'deconstruction of Christianity,' if it is ever possible, should therefore begin by untying itself from a Christian tradition of *destructio*."[18] The danger of the "almost impossible task" of deconstructing Christianity is that it is always at risk "of being exposed as mere Christian hyperbole."[19] How can this hyperbole be avoided? Can the danger be avoided, or is the threat of a Christian triumphalist victory necessarily imminent and always exceedingly possible?

I think that Nancy wagers that a stronger identification of deconstruction with Christianity would help undo Christianity and enable it to surpass itself, which is also means letting it go. But Derrida argues that this self-surpassing is at the same time indistinguishable from a Christian victory, which neither he nor Nancy desires. To be more specific, the victory that Nancy and Derrida desire to avoid is the victory of a triumphalist cultural and Western Christianity, what Kierkegaard calls Christendom, not Christianity as such, if there is such a thing.

Later in *On Touching*, Derrida turns to consider the work of Jean-Louis Chrétien, to show how close Chrétien's work is to Nancy's, and to acknowledge it as well as to attend to its deconstruction. Derrida exposes Chrétien's "rather complex strategy, which some might call devious."[20] On the one hand, Chrétien modernizes and Christianizes Aristotle via Thomas

Aquinas so that Aristotle can be oriented "toward a 'modern' Christian thinking of flesh." And on the other hand, Chrétien makes use of Aristotle, Aquinas, and also John of the Cross "in order to go against or beyond another 'modern' thinking of flesh that is phenomenological in kind."[21] These translations are also transubstantiations of God's hand and God's heart by means of God's Logos. Derrida says that Chrétien's "anthropotheo-*logical* thinking of flesh does not leave any spare room for a questioning of technics . . . nor of the animal, or rather animals, nor of the hominization process that produces what is termed the hand in 'everyday' language, nor of the possibility of prosthetics onto which spacing in general opens, and so forth."[22] That is, Chrétien makes room for a spacing, "but it is a finite place, which can be relieved, in the elevation of Logos and Incarnation."[23] The problem is that this Christian thinking of Incarnation resists any "irreducible finitude"; despite the rhetorical similarity to Nancy, Chrétien's phenomenology "would never lend itself to thought in what Nancy terms 'a finite thinking' for 'the sense of the world.'"[24]

Derrida worries whether Nancy can maintain the distance between his deconstruction of Christianity and Chrétien's Christianization of phenomenology. Derrida is aware that a straightforward embrace of the deconstruction of Christianity is a ruse because it will end up in a Christian victory that ultimately overcomes deconstruction, rather than the reverse. The problem, however, is that a simple opposition to Christianity is also insufficient because it gets caught in a similar trap where Christianity sublates this opposition and also wins. As Jacques Lacan says, "religion will always triumph" over psychoanalysis, philosophy, and many other things.[25] The problem is not Christianity as such, if such a thing exists, but the fact that Christianity as the absolute religion never loses. It always wins, even when it loses; when all is lost, it converts a loss into a gain, and a triumph over its pagan, Jewish and Islamic foes.

This is the Christianity that has become the modern West, as Walter Benjamin asserts in his unpublished fragment "Capitalism as Religion." According to Benjamin, Max Weber did not take his thesis on Protestantism and the spirit of capitalism far enough: "The Christianity of the Reformation period did not favour the growth of capitalism; instead it transformed itself into capitalism."[26] Christianity thrives on its own self-overcoming and its ability to subsume any and all alternatives into Christianity. This is the standard philosophical critique of Hegelian dialectics, that it is able to incorporate any and all negativity and emerge victorious. Lacan generalizes this triumphant Christianity to religion; religion gets to win every time there is a conflict. Modern capitalism emerges out of and draws on this self-surpassing nature of Christianity because it, too, is able to convert every loss into a gain, at least potentially. Every move against capitalism is appropriated within capitalism, so that it can never lose. Any specific person can lose or lose out; lose one's shirt, house, or investments; go bankrupt—but capitalism itself cannot lose because it lives off this gradient of win and loss in a way that structurally replicates the way that Christianity lives off of the opposition between salvation and damnation.

In the work of John D. Caputo and Martin Hägglund, two influential interpreters of Derrida's work on religion, we can see two alternative strategies to deal with Derrida's complex relation towards religion. For Caputo, most explicitly in his book *The Prayers and Tears of Jacques Derrida*, the challenge is to conceive religion and Christianity in Derridean terms without this triumphalist strain that emerges in globalized capitalism and reactionary fundamentalism. Caputo offers a religion *without* religion, even though he and Derrida both know that you never have one without the other. Religion as the affirmation of the promise, the primordial act of faith that underlies any performance, is one of the two sources of religion in "Faith and Knowledge," as we saw in the previous chapter.

In *Miracle and Machine*, which consists of a sustained reflection on "Faith and Knowledge," Michael Naas points out that Derrida shows a distinct preference to this act of faith over and above the other source, the experience of the sacred or holy as unscathed and indemnified, which leads to the structure of auto-immunity. Naas asserts Derrida's affirmation of faith as the fundamental social bond, but at the same time points out that Derrida wants to emphasize "the disruptive nature of this faith," which is a faith that always retains a hidden or secret side.[27] Caputo celebrates this secret faith of Derrida, this religion without religion that also means a Christianity without Christianism. Caputo says that "I would say that to save the name of faith, faith must be faith without faith, without the assurances of faith."[28]

Caputo does not collapse Derrida's tension between religion and religion, miracle and machine, or the two sources of religion, but he wagers on and elaborates a positive and affirmative view of what such a deconstructed Christianity would look like. Caputo claims that "Derrida's analysis of the gift makes possible another Christianity," not the same Christianity reconfirmed and redressed in postmodern garb.[29] He points out that for Derrida, "'God' is the name of the absolute secret, a placeholder for the secret that there is no secret truth."[30] God is not an object or a being but a potential event that resides in the structure of articulation as such, the secrecy that makes faith possible and that makes complete and certain belief impossible. Caputo later develops his vision of another Christianity in the form of a weak, deconstructed theology in *The Weakness of God*, as I will discuss in Chapter 6.[31]

Caputo acknowledges Derrida's atheism, but shows how it is still faithful. Derrida himself avows in an interview that "for my part, I believe in faith." And then he follows up with the claim that "as soon as one pronounces the word 'faith,' the equivocation is there, disastrous and deserted." We cannot avoid this equivocation or simply resolve it with an untroubled atheism or theism. "The religious," Derrida says, "in its equivocal relation to faith . . . is the equivocation in which we are."[32]

Derrida's equivocation and Caputo's affirmation are too much for Hägglund, however. In *Radical Atheism*, Hägglund argues that Derrida's affirmation of life and the living in its temporal becoming leads to a critique of immortality as the absolute immunity of life withdrawn from any exposure to death. Hägglund is correct to claim that Derrida does not affirm traditional faith religion in a conventional, otherworldly sense. According to Hägglund, "the spacing of time makes X possible while making it impossible for X to be in itself. Such spacing is quite incompatible with the religious ideal of absolute immunity."[33] There is no such thing as absolute immunity because spacing exposes every phenomenon to something that potentially compromises it. Hägglund understands Derrida's critique of the second source, the unscathed, which is incompatible with Derrida's emphasis on the temporization and spacing of language and life.

The problem with Hägglund's reading, however, is that it conflates Derrida's critique of the unscathed in terms of auto-immunity with his entire conception of religion, which includes belief, trust, or faith. Hägglund neglects the equivocation and partial affirmation of faith in Derrida's work, which leads Hägglund to view religion in entirely negative and atheistic terms, whereas Derrida himself is much more complex. In addition, Hägglund assimilates Caputo's interpretation of Derrida to a very simplistic and traditional view of religion, and fails to recognize that Caputo is elaborating upon Derrida's complex notion of faith.

Hägglund makes it clear that Derrida's affirmation of *salut* as a greeting or welcoming of the other is not a desire for salvation in any traditional sense. He points out that Derrida's "unconditional affirmation of life" involves saying "yes" to the survival of our finite temporal existence rather than hoping or yearning for some sort of religious infinity.[34] Hägglund is surely right to emphasize Derrida's focus on temporal finite existence rather than any sort of hope for immortality, and he offers a bracing corrective to some of the pious invocations of Derrida's thought and its ability to be subsumed into a more traditional religious framework. Hägglund is warning readers about the same misuse of deconstruction that Derrida warns Nancy about in terms of Chrétien's phenomenology in *On Touching*.

According to Hägglund, "what Derrida calls the impossible does not refer to something that is unattainable because of our human limitations, such as the kingdom of God. Derrida explicitly emphasizes that the impossible is not an inaccessible ideal; it is rather 'what is most undeniably real.'"[35] Hägglund points out that the spacing of time is what makes it possible for something to exist at all and at the same time makes it impossible for that something to be self-sufficient, or fully closed in on itself. This spacing of time is the time of life, and "it is quite incompatible with the religious ideal of absolute immunity."[36] I do not disagree with Hägglund here, but he misunderstands Caputo when he says that "Caputo's notion of the impossible . . . is the opposite of Derrida's."[37] Hägglund believes that Caputo figures Derrida's notion of the impossible as an ideal kingdom of God, which is incorrect.

How does Hägglund misread Caputo? Again, as already stated above, Hägglund assimilates Caputo's faithful reading of Derrida to only one of the two sources, the source of the unscathed and the autoimmune sacred. Hägglund argues that while Derrida distinguishes between faith and the unscathed, they "are usually conflated in the notion of religious faith, which is understood as the faith in an absolute Good that is safe from the corruption of evil."[38] But since Derrida critiques the possibility of any absolute immunity, Hägglund understands him to be entirely opposed to religion in both senses. And he assimilates Caputo to this stereotypical religious position, which is wrong. Caputo does not defend an indemnified sacred that is safe from deconstruction; he pursues a hopeful but risky path of a finite (but then deconstruction would also question the limits of any simple opposition between finite and infinite) faith. Hägglund claims that "Derrida undermines the religious ideal of absolute immunity, which informs both Caputo's and Kearney's reasoning."[39] But Caputo at least defends no such claim of absolute immunity, and does not endorse a God who is Good above and beyond all finite materiality.

Hägglund accuses religion of conflating faith with absolute immunity, and then assumes that because Caputo is defending or promoting religion that he is also committed to championing the absolutely unscathed source in the form of a pure God who is Good beyond being (which is closer to the position of Jean-Luc Marion than Caputo) rather than the finite source of religious belief. Derrida suggests that these two sources are in some way combined in what we call religion, and deconstruction attempts to tease them apart. Even if we cannot simply separate these two sources and cannot do away with one and keep the other, we can still think them distinctly and articulate them in important ways. Caputo risks confusion by defending religious faith in passionate terms, although this is a religion without (absolute, unscathed, immune) religion and faith without (secure, certain, guaranteed) faith. Hägglund thinks that he is more faithful to Derrida's atheism, but he gives up the equivocation that constantly pervades Derrida's writings about religion, an equivocation "in which we are."

For Caputo, Derrida's equivocation and careful affirmation becomes a more robust affirmation of religion without religion. Derrida always stresses the contestation, the ambivalence and ambiguous nature of religion, while Caputo wagers on the unconditional celebration of what Derrida carefully and hesitantly suggests. Caputo imagines a radical, ethical, self-contesting faith beyond any surety of the immunity of the unscathed, and this is not a betrayal or distortion of Derrida but a step along the way to the development of a full-throated Derridean theology, as we will see in Chapter 6. Derrida always resisted the term "theology" and distanced himself from it, and so did Caputo up until *The Weakness of God*. Radical, nonconfessional and unorthodox theology, however, offers Caputo a way to express Derrida's insights in more explicitly theological ways.

Although I think Hägglund misreads Caputo, I do not think that he literally misreads Derrida, even if he does betray the spirit of Derrida's philosophy to which Caputo remains faithful. They develop alternative strategies to safeguard deconstruction for or from religious faith, and they work hard and well to defend the coherence and relevance of their respective positions.[40] The problem with Hägglund's literalist reading is that it falls into a trap, which is similar to the trap in which Shylock finds himself in *Merchant of Venice*.

To help flesh out Derrida's complex attitude toward Christianity, I consider another essay by Derrida, one that is less commonly referenced when discussing his treatment of religion or the deconstruction of Christianity. In his essay "What Is a Relevant Translation?" Derrida discusses Shylock's impossible situation in Shakespeare's *Merchant of Venice* in a way that prefigures Hägglund's literalistic interpretation. In this essay, presented in 1998, published in French in 1999, and then translated into English in 2001, Derrida translates the first half of a sentence written by Shakespeare and uttered by Portia: "When mercy seasons justice, then must the Jew be merciful." Derrida focuses on the first phrase, which he translates into French as "*Quand le pardon relève la justice. . . .*"[41] Shylock holds fast to his oath to take a pound of flesh from near the heart of the merchant Antonio. His word is his bond, and he cannot forsake it. And yet, he is implored to be merciful as the doge of Venice has been and presumably will be merciful to him. Shylock eventually is forced to pardon Antonio and to convert to Christianity based on his stubborn insistence on the literal bond, and his lack of forgiveness ultimately leads to his death.

In his essay, Derrida focuses on the tension between a literal and a spiritual understanding of a word, an oath, a bond, and also a circumcision. Of course, he is not the first to treat this theme in *Merchant of Venice*, but it is interesting where Derrida ends up in the context of his own attitude toward religion and how it shows the limits of Hägglund's straightforwardly atheistic reading. Derrida states that "this impossible translation, this conversion . . . between the original, literal flesh and the monetary sign is not unrelated to the Jew Shylock's forced conversion to Christianity," because the Jew is often associated with the body and the letter whereas the Christian is seen on the side of a spiritualization of both word and flesh.[42] According to Derrida, the entire play proceeds

> as if the business of translation were first of all an Abrahamic matter between the Jew, the Christian, and the Muslim. And the *relève*, like the relevance I am prepared to discuss with you, will be precisely what happens to the flesh of the text, the body, the spoken body, and the translated body—when the letter is mourned to save the sense.[43]

The Jew is forced to pardon in a repetition of his own pardon, which is also a threat of death, just as his oath threatened Antonio with death. And this pardon is also a conversion to Christianity, a seasoning of Jewish justice with Christian mercy.

According to Derrida, this short sentence uttered by Portia—"When mercy seasons justice, then must the Jew be merciful"—"recapitulates the entire history of forgiveness, the entire history between the Jew and the Christian, the entire history of economics (*merces*, market, merchandise, *merci*, mercenary, wage, reward, literal or sublime) as a history of translation."[44] The Jew must be merciful, on pain of death. The Jew must forgive Christian violence, anti-Judaism and antisemitism, so that the Christian can forgive the Jew her own Jewishness and raise her up to the level of a Christian. Portia, a woman disguised as a man, admits that she has been paid as a "mercenary of gratitude, or mercy."[45] By mercifully commanding the Jew to be merciful, Portia "preconverts" Shylock to Christianity "by persuading him of the supposedly Christian interpretation that consists of interiorizing, spiritualizing, idealizing what among Jews (it is often said, at least, that this is a very powerful stereotype) will remain physical, external, liberal, devoted to a respect for the letter."[46]

According to Derrida, mercy in the play functions as a *relève*, a seasoning that makes justice more palatable and that also elevates it to a higher level. *Relève*, of course, is also Derrida's translation of Hegel's German term *Aufhebung*, which is "a dialectical movement of interiorization, interiorizing memory (*Erinnerung*) and sublimating spiritualization."[47] Mercy is viewed as a power of God that is available to humans to appropriate and use: "mercy *resembles* divine power at the moment when it elevates, preserves, and negates (*relève*) justice."[48] For Derrida, mercy instantiates itself at the level of the divine, in a "genesis of the divine, of the holy or the sacred, but also the site of pure translation."[49] Here is one of the two sources of religion, the holy or sacred, but at the same time it is a translation, so it is also a machinic iteration or repetition, a machine for making gods. And it is also the site of the link between the theological and the political, which Derrida wants to deconstruct, as we will consider in the next chapter.

This intersection of divine and human power in mercy constitutes Western Christian sovereignty, and the grandeur of this gesture is also a ruse of mercy, a sham of forgiveness that tricks and abuses—and eventually kills—Shylock. Derrida exposes and criticizes this Christian ruse of mercy in his essay. At the same time, Derrida has little mercy or sympathy for Shylock's victimization. He states that "I am not about to praise Shylock when he raises a hue and cry for his pound of flesh and insists on the literalness of the *bond*."[50] Why not defend Shylock? Why doesn't Derrida want to hold to the literality of the word, the oath, the bond or the body?

Derrida recognizes and expresses "the evil that can be through the Christian ruse as a discourse of mercy," but he still condemns Shylock's strategy of literalization. Derrida says at the end of the article: "I insist on the Christian dimension" of translation.[51] Why? Because otherwise, you are simply caught in its trap, which is where Hägglund ends up, because he is just a little too faithful to the letter of Derrida's work. Derrida insists on the spiritualization of interpretation, but then twists it into a work of mourning. Perhaps Caputo's own spiritualization of deconstruction fails to mourn Christianity explicitly enough, despite his emphasis on the tears of Derrida. So how do we survive Christianity? We mourn it, we let it die, and this is a relief (*relève*), in a sense.

At the end of his essay, Derrida suggests that we think relevance as memory and as mourning. He argues that translation guarantees *two* survivals, referring to Walter Benjamin: "by elevating the signifier to its meaning or value, all the while preserving the mournful and debt-laden memory of the singular body, the first body, the unique body that the translation thus elevates, preserves, and negates."[52] The translation thus preserves both the original literal word and the spiritualized double in a work of memory that is also a "travail of mourning." *Relève* is not just an elevation for Derrida; he gives the Hegelian *Aufhebung* a twist in its translation into French.

Relève is a relief that raises the body but also loses it, and in both losing and preserving it mourns rather than simply affirms and celebrates this spiritualization. The meaning of a translation is its relevance, and "the very concept, the very value of meaning, the meaning of meaning, the value of the preserved value originates in the mournful experience of translation, of its very possibility."[53] Derrida calls our attention to this travail, the work of mourning that haunts language in its iterability and translation.

Derrida claims that by simply opposing this *relève*, "Shylock delivers himself into the grasp of the Christian strategy, bound hand and foot."[54] You cannot fight Christianity directly, because you will always lose. Nancy wants to press deconstruction into Christianity to such an extent that Christianity will give itself up, surpass itself, which is what Christianity's essence has been from the beginning. But Derrida worries that this would be yet another Christian victory. And Hägglund wants to preserve the purity of his atheistic interpretation of Derrida uncontaminated by religion, which exhibits naïveté toward Derrida's complex thematic of auto-immunity as well as a betrayal of the "spirit" of deconstruction.

We can see how Derrida's reading of *Merchant of Venice* converges with Gil Anidjar's analysis of Christianity in *Blood: A Critique of Christianity*. In this book, Anidjar makes a striking connection between the economic circulation of capitalism in the modern world with the circulation of blood as discovered by William Harvey in 1628.[55] Blood is more than a metaphor; it is a complex, multifaceted material reality that spreads across our entire existence. Anidjar's book is a political theology because he traces the origin of this preoccupation with blood to Christianity, in particular the purity of blood. This obsession with the purity of Christ's—and then Christians'—blood eventually funds an entire racial economy. "Blood makes and marks difference," Anidjar writes, "an allegedly universal difference inscribed between bloods."[56] He paraphrases Schmitt later in the same paragraph by asserting that "*all significant concepts of the modern world are* liquidated *theological concepts*."[57] These concepts are liquidated, or liquefied, rather than simply secularized, which means that they still carry meaning and infect their bearers.

Everything begins with the blood of Christ. The sharing of Christ's blood in the communion ritual fuses a community together, a community that participates in the blood of Christ. According to Anidjar, "Christianity simultaneously invented the community of substance as the community of blood."[58] During the early history of Christianity, from the fourth to the eleventh century, it was considered a sin to shed blood, any blood, even of non-Christians. What changed is that around the late 1000s and early 1100s, during the so-called Papal Revolution, the Church declared that Christians should not kill other *Christians* rather than other humans. Urban II preached a crusade against Islam, claiming that "Christian blood, redeemed by the blood of Christ, has been shed" by infidels.[59] Anidjar relies on the work of Tomaž Mastnak, who says that from now on, "the church, which had long 'considered the shedding of blood as a source of pollution, now encouraged the shedding of blood—non-Christian blood—as a means of purification.'"[60]

Later on, in the 1449 Statutes on the Purity of Blood issued by the Spanish Inquisition, we can see "the institutionalized perception whereby Christians were deemed hematologically distinct from converts, the latter having failed to achieve Christianization by reason of their tainted blood."[61] Christian states such as Spain are called "vampire states" because they search out and destroy people whose blood is impure and non-Christian. The preoccupation with race and racism as represented by skin color is seen by Anidjar as an "intensification" of blood, where blood contains and carries the essence of race.

The origin of a certain ancient type of racism emerges with the origin of Christianity, which configures Jewish identity in ethnic terms in contrast with Christian universalism.[62] But this racial identity is not based on the notion of blood. Anidjar argues that in the Hebrew Bible, "there is *no difference between bloods*," and the Israelite or Jewish community is based on a sharing of "flesh and bone."[63] With the origins of Christianity in Paul, attention is paid to the flesh and blood of Christ, which become associated with the Eucharist ritual. Here blood enters the picture, although it is not yet associated with what we call race.

In the modern world, the significance of race emerges in connection with a purity of blood. Anidjar argues that "the earliest evidence appears to point to the Spanish and Portuguese empires, where purity of blood served in 'the forging of cleansed Spanish identity that referred both to national unity and to the overseas empire.'" In colonial Latin America, this idea of a purity of blood gradually became divorced from its religious significance and became a solely racial idea around 1700.[64] As Nelson Maldonado-Torres points out in an article on religion, conquest, and race in the modern world that is broadly compatible with Anidjar's study, "the Christian polemics and the discourse and practices surrounding the concept of the 'purity of the blood' are, in a manner of speaking, the anteroom to the modern racist discourse and practices that would be initiated with the arrival of Columbus in the Américas."[65]

For Anidjar, the implicit racial significance of religious difference between Christians on the one hand, and Jews and Muslims on the other, develops into the modern idea of peoples who are racially distinguished by their blood. Even if Anidjar exaggerates the significance of blood in the modern world, or his book downplays other elements, his examination is incredibly revealing, and it shows a deep connection between Christianity and contemporary racism and capitalism. Blood is a material reality as well as an object of political theology. Blood flows, out, during moments of exceptional violence, but our blood still matters even when it is less obvious or visible.

In his book, Anidjar reflects briefly on Shakespeare's play. He writes that "if *Merchant of Venice* teaches us so much about economic theology, it is first of all because it takes us to sea—war and commerce—while signalling to us that along with, and in the realm of, money, our minds are 'tossing on the ocean,' and merchants are 'like signors and rich burghers on the flood.'"[66] We are adrift on these flows, and the most important liquid will turn out to be blood, and the equivocation between money and Christian blood. Shylock is figured in the play as "the economic enemy whose association with blood and money is ultimately interrupted because and by way of blood, and more precisely, because and by way of Christian blood."[67] Anidjar offers a critique of the play that is similar to that of Derrida, although Derrida does not single out the theme of blood in *Merchant of Venice*.

For Derrida, the phrase delivered by Portia, "When mercy seasons justice, then must the Jew be merciful," is the most significant line of the play. For Anidjar, however, the key is Portia's judgment:

Take then thy bond. Take thou they pound of flesh.
But in cutting it, if thou dost shed
One drop of Christian blood, thy lands and goods
Are by the law of Venice confiscate
Unto the state of Venice.[68]

Shylock protests that in forbidding him his bond, the state of Venice is depriving him of his life, and this becomes prophecy. Anidjar draws attention to the circulation of life, money, and blood in the play and in our political-theological economy.

Shylock cannot shed a drop of Christian blood, even though he has claim to a pound of Antonio's flesh. This trap illustrates how Christianity functions because, as a Jew, Shylock is screwed no matter what he does. Derrida wants to avoid triggering the trap by opposing it directly; he wants to carefully explain it and if possible disengage it with as little bloodshed as possible. Anidjar risks a more direct critique of Christianity by focusing on its identification with blood and other liquids, including money. Anidjar renders Christian blood visible in its sacrificial, racial, and genocidal violence.

Whether or not one is Christian, there will be blood, which means that straightforward resistance is ultimately futile. You have to mourn Christianity, let it survive by dying (which it has always already done, by means of the Cross) and let it die by surviving, which it inevitably will, in its infinite translation. We will not survive, even if we want to, as ourselves. We have time, but not too much time, not all the time, and in certain respects we have no time. Space is available, and the "*khôra* of tomorrow . . . . *makes way perhaps*, but without the slightest generosity, neither human nor divine."[69] This is Good News, and we should mourn it, even if we should mourn as if not mourning—"for the whole frame of the world is passing away" (1 Cor 7: 30–31). This is the end of the world, but Derrida loves us and he is smiling.[70] Whew!

3

Political Theology Without Sovereignty

As we have already seen in previous chapters, in *Miracle and Machine: Jacques Derrida and the Two Sources of Religion, Science, and the Media*, Michael Naas offers a close reading of Jacques Derrida's 1996 essay "Faith and Knowledge: The Two Sources of Religion at the Limits of Reason Alone." Naas shows how these two sources of religion, the miraculous and the machinic, constitute what Derrida calls "religion" as faith and as the sacred. Furthermore, Naas traces these themes of miracle and machine across Derrida's corpus and shows how these questions related to religion animate Derrida's philosophy from the beginning to the end of his life. In addition, Naas intersperses his readings of Derrida on religion with some extraordinary connections to and reflections on Don Delillo's novel, *Underworld*.

I want to endorse Naas's reading of Derrida on religion, and his insistence on the two sources of Derrida's thinking about religion, and the role of religion in the contemporary world. In some ways Derrida is a "religious" thinker, but his religiosity is not simply equivalent to any specific religion. Naas explains that Derrida uses words like "faith," "God," and "messianicity" "in ways that court misunderstanding," but they are related to how Naas articulates the term "miracle" in his book. According to Naas, "Derrida says that we are called upon to believe every testimony—every claim to truth, every claim that one is telling the truth about what one knows, believes, or sees—as an 'extraordinary story' or a miracle."[1] The problem with the miracle, however, is that because of repetition it is always caught up in a kind of machine. "Derrida says that the machine is simply another way of speaking about calculation and repetition," argues Nass, "but about a calculation and repetition in relationship always to the incalculable and the unforeseeable."[2] This machinic repetition of the miracle both ruins the singularly miraculous quality of the miracle and also at the same time paradoxically makes it possible in the first place.

As soon as you have life, which is a miraculous event, you also have a kind of mechanics or what Arthur Bradley calls "originary technicity."[3] Religion, like life, conjoins the two elements of a singular respect for human existence with a kind of sacrificial repetition of this life that constitutes a mechanics, or a sort of death-in-life. This mechanical repetition "reproduces, with the regularity of a technique, the instance of the non-living or, if you prefer, of the dead in the living."[4] As Steven Shakespeare explains,

> The singular event requires an archive, if it is to survive and be read. This archive always has something "machinelike" about it: a capacity for iteration that exceeds or, rather, precedes the formation of conscious intentionality. This is the cut, the wounding of the present that ensures the possibility of survival. A kind of thinking of the cut, of the absolute—and of God—because this machinelike structure that presupposes itself, that causes itself, is strangely akin to God. Indeed Derrida understands the question of the technical as question of the theological.[5]

Religion in its two sources of the sacred and belief incorporates both the singular respect of life and the regularity of a non-living technique, such that the mechanical and theological converge, as Shakespeare points out. In "Faith and Knowledge," the theological machine is a "machine for making gods," as Derrida quotes Bergson from *The Two Sources of Morality and Religion*.[6] And these gods perpetrate miraculous acts, acts of sovereign exceptionality.

In amplifying Derrida's work by re-turning to and opening up this profoundly important text on religion, Naas helps us assess not only Derrida's philosophy but how it is that we can think and think about religion today. In his conclusion, Naas borrows an image in the form of a specific billboard sign from *Underworld* and claims that in their inextricable intersection, faith and knowledge, or miracle and machine, compose a sign that reads: "Space Available." Naas reads this sign as "an affirmation that, for the moment, there is still time, and in the words of the sign itself, which can be read as the translation of either a messianicity without messianism or *khôra*, still SPACE AVAILABLE."[7]

In this chapter, I focus more explicitly on a theme that Naas raises over the course of a few pages but does not fully elaborate, which is that of sovereignty. Derrida's religious writings are always connected to his political reflections, specifically the attempt to elaborate a "democracy to come" that would not be constituted in terms of sovereignty. According to Naas, the oneness or indivisibility of the sovereign is the legacy of particular religion—Christianity—and a particular political history—Western European, that needs to be contested and deconstructed. As Naas says,

> throughout the 1980s and 1990s and right up to his death, Derrida relentlessly pursued a kind of radical or originary secularism or secularity that constantly questions and criticizes the imposition of any particular religion or religious doctrine upon political concepts. Motivated in part by the analyses of Carl Schmitt, Derrida takes up the project of demonstrating the ontotheological origins of what at first appear to be modern secular concepts such as popular sovereignty, democracy, and religious tolerance.[8]

On this reading, a Christian political theology, of which we caught a glimpse in the previous chapter in Derrida's interpretation of *Merchant of Venice*, is entangled with sovereignty. Derrida appreciates the incisiveness of Schmitt's analysis of political theology in *Political Theology*, but he also criticizes and opposes Schmitt's conception of politics, most explicitly in *The Politics of Friendship*. My argument is that Derrida is working to deconstruct or dismantle the sovereignty that undergirds the political theology of Carl Schmitt. At the same time, Derrida is attentive to the inextricability of something like a theology in every politics, and the impossibility of completely separating questions of religion from questions of politics. Furthermore, I suggest that this thematic of the critique or deconstruction of sovereignty animates Derrida's text on "Faith and Knowledge," even if it is less explicitly foregrounded in the essay.

Sovereignty is what interlaces the theological and the political for Schmitt, which is why Derrida takes his distance from any explicit form of political theology.[9] The question of the deconstruction of sovereignty lies in what happens to the theological and the political if this link to sovereignty is undone. The main problem with sovereignty lies in its insistence on the "One," in Schmitt's case the one who decides on the exception. This oneness draws the political and the theological together into an antidemocratic machine of domination. For Derrida, there is no simple "one" who decides.

In *Two: The Machine of Political Theology and the Place of Thought*, Roberto Esposito lays out his understanding of the machine of political theology that encompasses the entire history of the West. According to Esposito, it is the specific relation of the two—the theological and the political—that constitutes a strange sort of unity. This unity or oneness is constituted by "a process of exclusionary assimilation." The Two are included in the One is a way such that "the imposition of one . . . seeks to eliminate the other."[10] Western humanity is caught within the workings of this machine. This machine of political theology is so effective and so inescapable because it comes to define human subjectivity, what it means to be a person. The dispositive of the person is constituted by the bond between subjectivity and subjugation, and this bond is internalized in the creation of human persons.[11] Esposito devotes his book to analyzing the history of the formation and development of this machine from ancient Rome and early Christianity up through modernity, and to finding ways to undo or render inoperable this machine.

The political theological machine functions by means of "an exclusionary selection of what it absorbs."[12] The Two become One by an assimilative exclusion of the other, which is never complete. This incompletion then drives the machine onward. In his book, despite its impressive scope, Esposito fails to consider Derrida's work in the context of his analysis of political theology, but I think that Derrida's thought is highly relevant to Esposito's analysis. Derrida attends to the deconstruction of the One into its constitutive Twoness, and opening up alternative ways of understanding and appropriating the history of Western Christian political theology.

In this chapter, I set up a context for Derrida's later work that stretches from *The Politics of Friendship* and includes *Specters of Marx* (and really begins with his 1989 essay "Force of Law: The Mystical Foundation of Authority") to his later book *Rogues* and his two-volume lecture course on *The Beast and the Sovereign*. These works all can be seen to pivot around the idea of the deconstruction of sovereignty and the possibility of undoing this "theologico-political" link. Here I briefly reflect on Schmitt's famous text, *Political Theology*, whose subtitle is "Four Chapters on the Concept of Sovereignty," along with his *The Concept of the Political*, and then consider some relevant passages from Derrida's texts mentioned earlier. Derrida's essay "Faith and Knowledge," along with Naas's reading of it, remains in the background.

I am claiming that the political emphasis of Derrida's later thought is consistent with the focus on religion in many of his works, because both of these make up what Esposito calls the machine of political theology. Derrida deconstructs sovereignty as a way to undermine the working of the machine, the paradoxical exclusionary inclusion of the Two in the One. Derrida does not explicitly treat the theme of political theology, but all of his works are related in some respect to this theme. This focus on the machine of political theology also opens up deconstruction from any literal association with writing. Religion and politics always interact and intersect, in powerful and problematic ways, and they do so not only in writing but also in broader transformative ways.

Derrida's philosophy has always been directed against the unity of the One. In his earlier work, his focus on dissemination is more literally a form of writing, a working against the presumed unity of speech. Dissemination involves a kind of double writing, as we can see in his essays "Dissemination" and "The Double Session" (Derrida's interpretation of Mallarmé), both of which are included in the book *Dissemination*.[13] Writing as dissemination exposes it to its otherness in a way that precludes gathering into a unity, and it punctures sovereign authority. But in his later works, Derrida's analyses are no longer configured specifically or generally in terms of writing.

In his later writings, Derrida draws closer to technics, or the machinic, and its connection to belief as a kind of promise that implies ethical responsibility as its condition of possibility and impossibility, as the underlying condition and context for thinking about the workings of religion and politics. In "Faith and Knowledge," for example, the context for reflecting on the two sources of religion is a more biopolitical conception of auto-immunity. The nature and role of religion in the contemporary world, which is one of "globalatinization," implies that we cannot simply dismiss or dispense with religion as a phenomenon. Religion constitutes a kind of auto-immune phenomenon that "is silently at work within every community, constituting it as such in its iterability, its heritage, its spectral tradition."[14] Religion is something intrinsic to community that threatens community with "a principle of self-destruction" in the form of violent fundamentalism. At the same time, this "self-contesting attestation keeps the auto-immune community alive, which is to say, open to something other and more than itself."[15] Here, religion is a form of auto-immunity that both protects and threatens the social body. Auto-immunity as a paradigm to think about religion invokes the machine, what constitutes the machinic, and what exceeds the machine as what makes it possible, what makes life worth living in its excess, its dignity, and its faith. Community and auto-immunity are not figures of *writing*; they constitute an opening to another form of conceptuality that Malabou calls plasticity. We could say that in Derrida, after writing recedes as a motor scheme, what comes to replace it can be viewed through the lens of technics and responsibility. I engage more with this theme in Chapter 7 by way of a more direct encounter with Malabou's philosophy.

My specific argument in this chapter is that, in his later work, Derrida is trying to sketch out a kind of political theology without sovereignty that would counter the political theology of Carl Schmitt, which is essentially tied to a form of sovereignty. Schmitt defines sovereignty in political-theological terms for Derrida, but Derrida wants to think about a radical politics of democracy that would not be defined as sovereign in Schmittian terms. Derrida accepts Schmitt's assertion that the political and the theological are both historically and structurally interrelated, but he imagines a future for political theory that is not tied to sovereignty. Because he associates Schmitt's conservative philosophy with the phrase "political theology," Derrida does not himself use this term. However, one way to read Derrida's later work is to see him as elaborating a political theology without sovereignty.

According to Geoffrey Bennington, a significant theme of Derrida's later work is "his call for an unconditionality without sovereignty." Sovereignty involves the effort to immunize the subject or sovereign from the risk of death that comes with the uncontainable uncertainty of an event. But for Derrida this unconditionality that is quasi-religious "involves exposure to the absolutely unexpected event as a condition of anything like 'life.'"[16] Unconditionality without sovereignty draws attention to the ways in which both politics and religion attempt to immunize themselves in dangerous and problematic ways. Derrida points out where and how both are exposed to the possibility of an event, and how this exposure is what makes them expressions of a living community even as it exposes them to death, the death of a machinic repetition. Political theology is a machine that names the crossing of religion with politics in our "globalatinized" world, even as it exposes the working of that machine to another conditionality that Derrida calls unconditional.

In his groundbreaking book *Political Theology*, published in 1922, Schmitt famously claims that "all significant concepts of the modern theory of the state are secularized theological concepts," and this is not only due to their "historical development," but also has to do with their "systematic structure."[17] This sentence opens the third chapter of the four chapters on the concept of sovereignty, and here Schmitt directly expresses the famous secularization hypothesis. In his analysis, Schmitt argues for a sociology of concepts, one that takes into account the genuine significance of ideas without simply arguing for or against one side or the other (secular vs. theological). He claims that during the modern period of European intellectual history, the battle against God represents an attack on transcendence for the sake of immanence, and this process results in modern atheism. In their attack on liberal atheism, conservative counter-revolutionary religious thinkers such as Donoso Cortés reveal what is importantly at stake in this transformation. It's not simply theism vs. atheism; the problem for Schmitt is that liberal democracy represents an "onslaught against the political."[18]

The revolt against God ends up destroying not only religion but politics as well. Why? Because our understanding of God is tied to a modern conception of sovereignty, and the basis of sovereignty is its personalistic decision-making capacity. The first sentence of *Political Theology* reads: "Sovereign is he who decides on the exception."[19] A decision has to be a sovereign decision to be a true political decision. Schmitt criticizes the degeneration of modern law and economics to the point where it attempts to eradicate the need for any decision. This elimination of the decision is both impossible and undesirable. At the end of the book, he argues that

> Whereas, on the one hand, the political vanishes into the economic or technical-organizational, on the other hand the political dissolves into the everlasting discussion of cultural and philosophical-historical commonplaces, which, by aesthetic characterization, identify and accept an epoch as classical, romantic, or baroque. The core of the political idea, the exacting moral decision, is evaded in both.[20]

Sovereignty is tied to the possibility of a political decision, which is an exacting moral decision. In overthrowing the sovereignty of God, humans are attempting to get rid of sovereignty altogether. But this effort eliminates all politics, which Schmitt wants to hold onto. Politics requires the sovereign decision in order to be morally exacting, and there is always a residual structure of the theological in every moral-political decision.

In a book written later in the 1920s, *The Concept of the Political*, Schmitt further clarifies what he means by the political, which is the famous distinction between friend and enemy.[21] It is the ability to make such a distinction that renders an action political or morally exacting in its highest sense. And this ability to make such a clear and absolute distinction is precisely what Derrida contests in *The Politics of Friendship*. Derrida suggests that an opposition between friend and enemy of the sort that Schmitt sets up necessarily deconstructs. There is an aporia in the event of a political decision that ruins sovereignty at the same time it upholds it. For Schmitt, politics depends on the decision concerning who is a friend and who is an enemy, and the enemy is then the object of political hatred and war. But Derrida notes that "the fundamentally Christian politics" of Schmitt, is only possible when thought under a "Christian metaphysics of subjectivity."[22]

According to Derrida, the sovereign decision is ruined from within by an aporia of responsibility. He argues that "*a theory of the subject*" such as

Schmitt's "*is incapable of accounting for the slightest decision.*"[23] This is because every active or autonomous decision is also exposed to a passive decision as its necessary condition. "The passive decision," Derrida writes, as the "condition of the event, is always in me, structurally, another event, a rending decision as the decision of the other."[24] Schmitt's political theology cannot account for this other decision that inhabits and wrecks the sovereign decision that decides on the exception. Derrida says that "the decision is not only always exceptional, *it makes an exception of/for me.*"[25] The sovereign decision makes an exception for me, and I have the power to decide, but this decision "exonerates from no responsibility." Responsibility, which lies at the heart of Derrida's faith, as one of the two sources of religion, means that I am "responsible from myself before the other, I am first of all and also responsible *for the other before the other.*"[26] The temptation is to read this responsibility as a moral exhortation, but for Derrida it is a much deeper and more structural situation that he develops from the philosophy of Levinas, and this responsibility constitutes the self as self. As the Indian philosopher Saitya Brata Das explains, Derrida "speaks of a messianic exception which for him is the true exception, and which is different from the exceptionality of the sovereign in respect to law."[27] The true, messianic exception is the general case of exception that makes us responsible as subjects for an other. This exceptionality both constitutes and ruins responsibility because it makes us responsible, but it also makes us irresponsible, because we can never do justice to the other, not even to ourselves as other.

Derrida wants to avoid or overcome the Christian metaphysics and politics of Schmitt, but he understands that there is an irreducible religious element to all politics and all philosophy, so he adopts the term "messianic," or a messianicity without messianism as he articulates it in *Specters of Marx*. "The messianic appeal," he writes, "belongs properly to a universal structure, to that irreducible movement of the historical opening to the future, therefore to experience itself and to language."[28] This universal structure of messianicity is contrasted with any determinate messianism as religion. It is a "religion without religion," as Derrida says in *The Gift of Death*. The human being is marked by this exceptional messianicity and this unavoidable responsibility, divided and shared as she is between a pure animal and an absolute sovereign.

In his two-volume lecture course on *The Beast and the Sovereign*, Derrida tracks these themes of sovereignty and animality through Western literary and philosophical discourse. In Volume I, he claims that he is working against Schmitt's "cunning intensification of the political" that relies on a conception of evil by which to judge the political enemy, but he affirms the necessity for "an other politicization" rather than a depoliticization, "and therefore another concept of the political."[29] This other concept of the political is based on a deconstruction of sovereignty. For Derrida, sovereignty is not indivisible; it is divisible and divided. And "a divisible sovereignty is no longer a sovereignty, a sovereignty worthy of the name, i.e. pure and unconditional."[30] It's not enough to simply change subjects or exchange sovereigns from God to monarch to people: "the sovereignty of the people or of the nation merely inaugurates a new form of the same fundamental structure."[31] The deconstruction of sovereignty issues not in a liberal democracy, however politicized. Derrida affirms a more radical form of democracy, a democracy *to come*.

Schmitt takes aim, along with many other conservative and radical critics, at liberal democracy. He offers a scornful dismissal of liberalism in *Political Theology* to the effect that "liberalism . . . existed . . . only in that short period in which it was possible to answer the question 'Christ or Barabbas?' with a proposal to adjourn or to appoint a commission of investigation."[32] In a similar vein, Alain Badiou rails against contemporary parliamentary democracies and their inability to recognize radical evil.[33] Derrida does not want to defend existing liberal or parliamentary democracy, but he does want to save the name of democracy, and to inflect it in more futural terms.

In *Rogues*, Derrida shows how the question of democracy is still necessarily caught up with the question of God, even if he wants to think both democracy and divinity without sovereignty. Most contemporary reflections on democracy see it as tied to a form of sovereignty. He says that "now, democracy would be precisely this, a force (*kratos*) a force in the form of a sovereign authority . . . , and thus the power and ipseity of the people (*dēmos*)."[34] Is it possible to have democracy without sovereignty, and if so, what would that mean?

The problem is that any sovereignty necessarily involves the One, the authority of the One who acts and decides, whether the sovereign is God or King or People. Even if democratic sovereignty relies on the sovereignty of the people, it retains "the sovereignty of the One . . . above and beyond the dispersion of the plural."[35] A state-form that relies on any form of sovereignty is in some respects a rogue state, because the force of sovereign authority ultimately comes down to the "reason of the strongest."[36] For Derrida, an affirmation of democracy would have to mean a dispersal of the One, a sending off and away of sovereignty, which means that democracy is never simply present but always also futural, "to come." This futurity at the heart of the present is also a kind of messianicity insofar as it is never fully present, but also gestured toward as in to the arrival of an apparition, or a ghost.

A post-sovereign democracy would have to be plural; it would have to pluralize and thus dislocate or deconstruct sovereignty, especially the sovereignty of the One. Schmitt's political theology affirms the sovereign act as a way to maintain the integrity and seriousness of politics. But Derrida attempts to think politics beyond or without sovereignty, without the sovereign one that possesses the authority to make a genuine political decision. This attempt destroys realist power politics, but it remains a vital hope for any person or people who strive for justice. A politics based on justice in the Derridean sense has to affirm democracy, not as an actual state of affairs, but as Derrida explains, "the democracy to come would be like the *khôra* of the political."[37] *Khôra* is the space available for political negotiation, that resists the closing in on itself of a sovereign One.

Derrida wonders whether his understanding of democracy and his advocating of a "democracy to-come" "might not lead back to or be reducible to some unavowed theologism."[38] Politics cannot be completely dissociated from religion, and there is a religious messianicity that haunts all our politics, even if we can keep politics from degenerating into this or that messianism. He asks, "the democracy to come, will this be a god to come? Or more than one? Will this be the name to come of a god or of democracy? Utopia? Prayer? Pious wish? Oath? Or something else altogether?"[39] Derrida suggests that democracy has some connection to Heidegger's late thinking about God, as expressed in the posthumous interview, "Only a God Can Save Us," even though Heidegger resists such a connection. By reflecting carefully about his own complicated relationship to Heidegger, Derrida concludes his essay by suggesting that his own (Derrida's) understanding of democracy is tied to the idea "of a god without sovereignty," even though "nothing is less sure than his coming, to be sure."[40] Nothing is less sure than a god who could save us, even if that is the only thing that can save us now. And nothing is less sure than a democracy to come, even if that is the only hope for a responsible politics.[41]

Just as Derrida wonders whether Heidegger can avoid writing a theology, we are justified in asking whether Derrida, in fact, has a political theology. To be sure, he does not use that name, and I think that it mainly due to his desire to distance himself from Schmitt's political theology. But the question remains: Is there only one political theology, that of Schmitt, or is there more than one? This is the question, and the promise, of Jeffrey W. Robbins's book *Radical Democracy and Political Theology*, which seeks to develop a political theology that would not be Schmittian. Robbins relies on the work of Michael Hardt and Antonio Negri to articulate a political theology of the multitude that constitutes a radical form of democracy rather than a unitary sovereign power.

Robbins cites Derrida's criticism of Schmitt in *The Politics of Friendship*, and claims that this encounter "leaves us with the question of whether a political theology might provide the necessary supplement to contemporary democratic theory and practice without falling prey to the same exclusive logic as Schmitt."[42] I want to suggest that this is partly what Derrida is trying to do, but he avoids the phrase "political theology" for the same reason that so many theorists do, because of the long shadow cast on European radical thought by Carl Schmitt. According to Antonio Negri, whose philosophy Robbins draws upon heavily in his book, there is only "just one political theology, the one at whose opposite ends stand Bodin and Stalin, with Carl Schmitt occupying a slot somewhere in between."[43] If Negri is right, then Robbins is wrong, and there is no possibility for any political theology without sovereignty because sovereignty *defines* political theology.[44]

Derrida opposes political theology in name, but he also recognizes how religion cannot be exorcised from politics, from "Force of Law" to his last writings. Instead of political theology, he works with the quasi-religious category of messianicity, a messianicity without messianism, because it is a "weaker," or less sovereign, power. In a 1994 essay, "Taking a Stand for Algeria," Derrida takes "a stand for the effective dissociation of the political and the theological." He states that "our idea of democracy implies a separation between the state and the religious powers, that is, a radical religious neutrality and a faultless tolerance which would not only set the sense of belonging to religions, cults, and thus also cultures and languages, away from the reach of any terror . . . but also protects the practices of faith, and, in this instance, the freedom of discussion and interpretation within each religion."[45] The religious power that the theological represents, when aligned with the political force of the state, produces terror.

Does political theology necessarily imply this sovereign power and therefore a rogue state with its concomitant violence and terror? Perhaps. But what if the theological itself could be weakened, or viewed as non-sovereign, as John D. Caputo has been insisting for years, following Derrida? In *The Weakness of God*, Caputo articulates a thinking of divinity without sovereignty. Here God names the unconditional solicitation of an event rather than

a being who is there, an entity trapped in being, even a super-being up there, up above the world, who physically powers and causes it, who made it and occasionally intervenes upon its day-to-day activities to tweak things for the better in response to a steady stream of solicitations from down below.[46]

In his later book *The Insistence of God*, Caputo argues for a theology of "perhaps" that is tied to a weak non-sovereign force rather than traditional sovereign power. As he explains, "the 'perhaps' of which I speak here does not belong to the 'strong' or sovereign order of presence, power, principle, essence, actuality, knowledge, or belief."[47] *Perhaps* means a way to say yes to the future while affirming the "chance of the event," the chance that the event might not happen as well as the chance that the event might not be what we want, or even good; it might be the worst. I will return to Caputo's interpretation of Derrida and the development of his own form of radical theology later in chapter 6.

According to Derrida, as he argues in *Rogues*, sovereignty is a circular movement; he says that it forms a kind of merry-go-round, a rotary motion that draws power in toward itself and then distributes this same power that it has appropriated for itself back out to the rest of us who are nonsovereign. Now, today, much of this sovereignty has become invisible; we exist in what Deleuze calls a society of control where most of the time we consist of relay points for the distribution and redistribution of sovereignty. The focal point or center of sovereignty is invisible, it appears not to exist, but it functions all-the-more smoothly despite this inexistence. The center of sovereignty is everywhere and nowhere, and its operation seems ubiquitous and invisible, like the spirit of God. How do we track down the machinations of sovereignty so that we can understand and perhaps even disable it, render it inoperable?

Is what Derrida calls the unconditional or the messianic a new form of sovereignty, precisely because it is undeconstructable? It is a pure force and it keeps us awaiting, expecting the unexpected and the unbelievable, like the end of the nation-state or the collapse of global capitalism. Is our faith in God or in democracy precisely what keeps us from resisting or is it what gives us the power to fight back, or is it instead what gives us the strength to resist actually the onslaught of actually existing corporate capitalism? These questions are literally undecidable but nevertheless imperative to think and to think through.

Political theology reemerges as a discourse when modern liberalism enters into a state of crisis, as foreshadowed by Schmitt. Contemporary political theology in the philosophy of Derrida and Giorgio Agamben attends to the fundamental crisis of sovereignty of the modern nation-state. Nations see their sovereignty superceded by flows of money, debt, and energy. In postcolonial terms, the challenge has been to track the emergence of "a global order of empire without colonies," as Partha Chatterjee puts it.[48] In his book *The Black Hole of Empire*, Chatterjee adopts Schmitt's basic theory of sovereignty and applies it to imperial power. He says that "the imperial prerogative as the power to declare the exception is useful" for making sense of the contemporary world order, not only in economic terms but also as a pedagogical project.[49] In a similar vein, Achille Mbembe recounts the policies during the 1980s and 1990s that have alienated the political sovereignty of African states and "created the conditions for a privatization of this sovereignty."[50] The deregulation and primacy of the market in neoliberal capitalism coincides with the proliferation and intensification of "private military, paramilitary, or jurisdictional organizations."[51]

As the economies of the United States and Europe have weakened over the past few years, Chatterjee now worries that "the asymmetry between the economic troubles of the Western powers and their overwhelming military superiority could well open the field for a populist resurgence of imperialism, not unlike what was seen in the late nineteenth century."[52] Furthermore, Thomas Picketty's economic analysis in *Capital in the Twenty-First Century* has shown that the concentration of capital in Europe in the early twenty-first century is very similar to that of the late nineteenth century.[53] We could call this situation a neo-liberalism that devolves into neo-imperialism, a drastic last resort to shore up the power of Western-style capitalism.

We do not need to endorse Schmitt's conclusions to make use of his concepts, as Derrida has shown. Political theology is not solely about religion, but it uses the situation of religion in the contemporary world as a way to think about sovereignty. Sovereignty for Schmitt and for Chatterjee means the ability to decide on the exception, but for Derrida, sovereignty deconstructs. For Schmitt, sovereignty is based on a unity, a unified decision-making power. Today, sovereignty is divided such that there is no absolute sovereign, even if there exist what appear to be sovereign decisions. Monarchical sovereignty is essentially related to monotheism, where the one king represents and decides in the place of the One God. For Derrida and other theorists, however, sovereignty is dispersed, divided, or shared such that there is no One Decider. The question is then whether there remains sovereignty. Who decides what matters today—the nation, the bank, the market, the army, the people, the consumer? What about the sovereignty of the planet?

The terminal crisis of capitalism is the ecological limit to growth. Technological developments have allowed us to evade the consequences of over-population and over-exploitation of natural resources, but these developments have taken place with and within the context of transformations of energy use. The extraordinary material achievements of human civilization over the last couple of centuries, however unevenly distributed, are based on cheap energy, primarily fossil fuels. Alternative energy sources cannot provide the storage, scale, or EROEI that oil and gas, and to a certain extent coal, provide. In addition, we are reaching real limits to growth in terms of atmospheric absorption capacity, rare earth metals, arable farm land, and fresh water. As Michael T. Klare explains in *The Race for What's Left*, we face an extraordinary combination of factors:

> A lack of any unexplored resource reserves beyond those now being eyed for development; the sudden emergence of rapacious new consumers; technical and environmental limitations on the exploitation of new deposits; and the devastating effects of climate change. In many cases, the commodities procured during this new round of extraction will represent the final supplies of their type.[54] (Klare, 18)

Capitalism requires growth, and since the 1970s, growth has slowed and resources have become more expensive. The transition to neo-liberalism and the increasing concentration of wealth is what happens when you cannot grow in absolute terms, but only in relative terms.

We need to think about capitalism and neoliberalism, religion and politics in an ecological context. We need to attend to the complex material flows of energy, power, food, climate, money, and blood, as seen in Gil Anidjar's book *Blood*, considered in the previous chapter. Derrida's philosophy matters in these contexts, even if his work has been canalized into more textual discourses. In *The Beast and the Sovereign*, Derrida reflects on the location of the human between the animal and the divine, and he attends to the limits of humanity insofar as humanity is implicated in animality and divinity. In modern liberalism, the normal human being exists under the sway of the law, but "sovereign and beast seem to have in common their being-outside-the-law."[55]

Who makes an authentic decision today? God, the autonomous self, the anonymous media, the nation, the market, the brain, or the earth? And what criteria can be applied to determine whether this decision is for the best, or at least for the better? Derrida's faith is that there are not only multiple decisions but always more than just one decider, and that this irreducible plurality at the heart of decision prevents the worst concentration of power in the hands of the One. It's the only hope for democracy, if there is such a thing.

What about the sovereignty of death? At the end of Volume II of *The Beast and the Sovereign*, Derrida returns to his decisive confrontation with Heidegger, armed with a phrase from a poem by Paul Celan and a sustained engagement with the novel *Robinson Crusoe*. The last line of Celan's poem "Vast, Glowing Vault" is: "the world is gone, I must carry you." Derrida reads this line and this poem carefully in his essay "Rams," and I will discuss this in the following chapter. Here I want to note how Derrida relates the carrying or the bearing—the German word is *tragen*—to a kind of sovereignty in Heidegger.

Derrida says that it is not clear whether the word *tragen* refers more to an experience of mourning the dead person, "or toward the child to be born and still carried by its mother or even toward the poem and the poet himself."[56] Derrida tends to associate the notion of the world as gone (*fort* in German) with death. Death is the end of the world, and here the world itself is gone, but even in the fort of the world, *ich muss dich tragen*, I must bear or carry you. This thinking of *tragen* is related to a thinking of sovereignty in Heidegger's philosophy. Sovereignty in German is *Souveränität*, but Derrida focuses on Heidegger's word *Walten*, a word that means rule, reign, or governing. Derrida follows Heidegger in bringing together *tragen* and *Walten* around the experience of death.

Sovereignty has to do with the ontological difference, the difference between Being and beings. Being and beings are related in an *Austrag*, or conciliation, that attests to the sovereign force of difference.[57] At the end of the seminar, Derrida asserts that *Walten* as ruling power refers to "the event, the origin, the power, the force, the source, the movement, the process, the meaning, etc.—whatever you like—of the ontological difference."[58] The relationship between Being and beings constitutes a world, but this world is decomposed in death. Heidegger says that animals cannot die in an authentic manner; only Dasein can do so. Derrida questions this assumption, as he deconstructs the opposition between beast and sovereign in his seminar. Derrida claims that "no one in the world will deny, not even Heidegger, then, that both types of living being [beast and sovereign] cease living, find death."[59] One can always find death as a living being, and all living beings exist as cohabitants of a "common habitat, whether one calls it the earth (including sky and sea) or else the world as world of life-death."[60]

We presuppose a common world as the envelope of this life-death that we cohabit, but this common world is also in a state of radical dissemination: "perhaps there is no world."[61] Perhaps instead we share an archipelago of islands that are radically un-shareable. "There is no world, there are only islands."[62] The radical un-shareability of worlds means that there is no tenable difference between the beast and the sovereign, even if one makes himself like Robinson Crusoe the master of an island and subdues all the beasts.

In this final seminar, delivered shortly before his death, Derrida argues that despite a certain superficial similarity with Schmitt, Heidegger's understanding of sovereignty is in fact radically different. Heidegger uses the word *Walten* as a kind of sovereign force, but his usage indicates an "excess of sovereignty" that undoes "the limits of the theological-political. And the excess of sovereignty would nullify the meaning of sovereignty."[63] Derrida says that for Heidegger Dasein is gripped by, exposed to an originary violence. This violence is a kind of "effractive departure from self in order violently to break open, to capture, to tame" beings and to treat them as beings, "*as* sea, *as* earth, *as* animal."[64] This violence that is proper to the human allows us to discover and make use of beings as such.

The violence of Dasein as it is constituted in and by the rift/accord/difference or *Austrag* between Being and beings gives humanity its sovereignty over the rest of the world. Derrida presses against the limits of this violence and the possibility of thinking about the world as a world, and I will come back to this in the next couple chapters. Here, in a striking conclusion Derrida cites Heidegger's argument that "there is only one thing against which all violence-doing, violent action, violent activity, immediately shatters," and "it is death."[65] Death is an absolute limit that shatters the violence of human sovereignty, even as it reigns in a hyper-sovereign way.

The question Derrida asks is, how do we know who can die? And this decision on death is a form of sovereignty, even though at the same time it exceeds any form of sovereignty that we know. Here the sovereignty of death exceeds the sovereignty of any sovereign and deconstructs the link between the political and the theological. This is a genuine "gift of death." Death is the limit of sovereign violence, even as it wields its own "sovereign" violence over life. Death removes the traditional sovereignty from political theology, the conjunction of the beast, and the sovereign that animates Western politics and metaphysics. This non-sovereign sovereignty of death delivers us to another scene of political theology. Our shared death is precisely un-shareable, but it is quintessentially democratic.

In the following chapter, I return to Derrida's reading of Celan's poem in connection to Heidegger as he discusses it in "Rams." In "Rams," Derrida reflects on Gadamer's death and Gadamer's hermeneutics, but he ends up as always with a consideration of and confrontation with Heidegger. The ram in the poem and in the biblical tradition, particularly the Aqedah, becomes a figure of death as well as a figure of divinity.

4

Interrupting Heidegger with a Ram
Derrida's Reading of Celan

This chapter is a kind of interruption as well as a continuation of my readings of Derrida and religion. We saw in the previous chapter how Derrida
considers Celan's phrase "The world is gone, I must carry you" along with
his reflections on *Robinson Crusoe* and Heidegger in *The Beast and the Sovereign*, Volume II. My understanding of Derrida's reading of Celan here
is not a mere focus on poetics in the form of writing, but a much broader
perspective on Derrida's thought. Of course, this was always the conception of writing for Derrida, but it becomes more and more evident in his
later work. My argument is that Celan's poetry gives Jacques Derrida something incredibly important, which is a kind of writing and a kind of
thinking that allows him to interrupt Martin Heidegger, who is the predominant philosophical voice in Derrida's work.

The primal scene for this interruption at the source concerns the famous
encounter between Celan and Heidegger at Heidegger's home in Todtnauberg in 1967, which Celan memorialized in a poem with the same name.
Heidegger attended a reading by Celan in Freiburg and invited him to his
home in the Black Mountains of southern Germany where Celan signed his
guestbook and they took a short walk, after which Celan was driven back
to Freiburg. Celan published his poem "Todtnauberg" in 1968, and it is
an incredibly ambivalent testimony of this fraught encounter.[1]

The longest stanza of "Todtnauberg" references Heidegger's guest book,
which Celan signed, and he asks who signed the book before he did, before mentioning

the line inscribed
in that book about
a hope, today,
of a thinking man's
coming
word
in the heart,[2]

This "hope of a thinking man's coming word in the heart" indicates Celan's desire for Heidegger to acknowledge his Nazi activities in the 1930s and take responsibility for his refusal to apologize for them after the war. The poem indicates Celan's disappointment, since on the drive back there was

coarse stuff, later, clear
in passing,[3]

on which the driver listened in. The end of the poem offers an image of the walk Celan took with Heidegger, which ended up being a damp slog over half-trodden logs rather than a fresh stroll along a path into a clearing: "the half- / trodden fascine / walks over the high moors, / dampness, / much."[4] The English translator Michael Hamburger translates *Knüppel*, which generally means sticks, as fascine, to indicate the connection with fascism as a bundle of sticks, although this might be a little overdetermined. At the same time, the last line of the poem, "dampness, / much" (*Feuchtes, / viel*) does not suggest any sort of resolution or exhilaration, but rather a desultory culmination of their encounter.

For Celan, this encounter was an unsatisfying one as expressed in the poem, although the German philosopher and student of Heidegger Hans-Georg Gadamer puts a much more positive spin on it, celebrating the fact that Celan and Heidegger could have this encounter despite their pasts, and revealing that Heidegger greatly enjoyed the poem "Todtnauberg" and had it framed. According to Gadamer, Celan was "among the pilgrims who made their way to Todtnauberg, and from his meeting with the thinker a poem came to be."[5] Gadamer reproduces the poem at the end of his reflections on Heidegger in his book *Philosophical Apprenticeships*. He mentions the line of hope, but ascribes it to Celan's desire to have a meaningful encounter rather than an expectation of Heidegger to take responsibility, and Gadamer then says that they walked "across soft meadows, both alone, like the individually standing flowers—the orchis and the orchid."[6] It is then only on the way home that Celan realized that what Heidegger said still seemed crude, but the walk itself was an act of breathless daring. This interpretation follows the chronology of the meeting, but not the poem, which ends with the walk. And the walk was not a daring engagement between two solitary people, a thinker and a poet, but rather a damp trudge that went nowhere. In the poem, the crudeness precedes the walk and leads up to it. In his text, Gadamer idealizes Celan's poem as well as Celan's encounter with Heidegger.[7]

Derrida clearly remembers Gadamer's misunderstanding of Celan's encounter with Heidegger when he is invited to say something about Gadamer after Gadamer's death in 2002. Derrida's essay, published in French in 2003, is entitled "Rams: Uninterrupted Dialogue—Between Two Infinities, the Poem."[8] In this powerful essay, one of Derrida's last works before his own death in 2004, Derrida reflects on Gadamer's life and thought, as well as Gadamer's understanding of Celan. Derrida does not directly discuss "Todtnauberg," although he does barely mention it at the end of the essay. Instead, he analyzes a poem by Celan called "Vast, Glowing Vault." At the end of his essay, Derrida concludes with some reflections about Heidegger, as he so often does.

This is a complex encounter among four extraordinary thinkers: Derrida, Celan, Gadamer and Heidegger. In the background of Derrida's essay lies another encounter, a broken-off or interrupted encounter between Derrida and Gadamer in Paris in 1981, where Derrida seemed to rudely refuse to engage Gadamer in dialogue.[9] So Derrida is reflecting in "Rams" on his own previous non-encounter with Gadamer, Gadamer's misunderstanding of Celan, and Celan's ambivalent encounter with Heidegger. Furthermore, there is Derrida's own appreciative but complicated relationship with Celan, who taught German at the École Normal Supérieure in the 1960s, and was introduced to Derrida by Peter Szondi in 1968. Derrida explains in an interview that "Celan's presence was, like his whole being and all his gestures, extremely discreet, elliptical, and self-effacing. This explains, at least in part, why there was no exchange between us, although for some years I was his colleague."[10] Even after they met, Derrida says, "a series of meetings can be dated, always brief, silent, on his part as on mine. The silence was his as much as mine."[11] This silence prevailed over any kind of dialogue, conversation, or interaction, and it was cut off in a way by Celan's suicide in 1970, which prompted Derrida to work to recover the significance of these brief but powerful memories. He says that "with regard to Celan, the image that comes to mind is a meteor, an interrupted blaze of light, a sort of caesura, a very brief moment leaving behind a trail of sparks that I try to recover through his texts."[12]

One of Derrida's earlier essays also focuses on Celan, although I am not going to analyze it here except to acknowledge it in very general terms and to indicate one specific reference. In "Shibboleth: For Paul Celan," Derrida reads Celan's poem by the same name and reflects on the concepts of circumcision, date, and language that "Shibboleth" raises in its reference to the month of February. Here is the stanza Derrida focuses on:

Heart:
here too reveal what you are,
here, in the midst of the market.
Call the shibboleth, call it out
into your alien homeland:
February. *No pasarán.*

Derrida claims that the last line of this stanza refers to February 1936, when the Spanish Republicans won an electoral victory against fascism, a "no" to Franco, Mussolini, and Hitler. And three years later, during the siege of Madrid, "*no pasarán* was a *shibboleth* for the Republican people, for their allies, for the International Brigades."[13] In another poem, "As One," Celan specifically mentions February 13, and Derrida suggests that one of the references here is to February 13, 1962, and a massive march in Paris for the victims of a massacre near the end of the Algerian war.[14] Derrida cites this second poem because Celan repeats the same phrase from "Shibboleth," *no pasarán*, right after he refers to the "Peuple de Paris." Here Derrida reads "Shibboleth" along with the poem "As One" in a meditation on the nature of language and the circumcision of language because it is the mispronunciation of the word shibboleth as sibboleth that gives away the enemy, and defines the border between nations and peoples, originally in the encounter between the Ephraimites and the Gileadites as related in the Book of Judges.

As Derrida claims in his reading of Celan, "shibboleth is a circumcised word" that is an "unpronounceable name for some."[15] There is a sense in which all of Celan's poetry functions as a kind of shibboleth because it is so difficult to read and understand, despite Celan's insistence that his work is "ganz und gar nicht hermetisch"—absolutely not hermetic.[16] Another reason Derrida turns from the poem "Shibboleth" to the poem "As One" lies in the final line of the latter poem: "*Freide den Hütten!*" Peace to the cottages or huts, a line "whose terrible irony must surely aim at someone."[17] So Heidegger, who ends up isolated in his hut in the Black Forest, appears, however indirectly, in Derrida's substantial essay on Celan, "Shibboleth." And he reappears, more explicitly, at the end of Derrida's later essay "Rams."

At the center of "Rams" lies a poem by Celan, and Derrida's subtitle suggests that the poem lies between two infinities. I will come to the poem in a little bit, but let me suggest that these two infinities progress in a kind of parallel line, without meeting or touching along their trajectories. I will not hesitate to name these two infinities for Derrida: One name is the infinite philosophical thought of being of Heidegger, and the other is the infinite ethical thought of Emmanuel Levinas, which takes place in some respects "otherwise than being." Derrida has written and thought about Levinas and Heidegger more than all the others, and much secondary work has been done on these two powerful influences on his work. In fact, it could be argued that the most crucial essay of all of Derrida's corpus is his incredible essay on "Violence and Metaphysics," which appears in *Writing and Difference* and has the subtitle, "An Essay on the Thought of Emmanuel Levinas."[18] Although Levinas does not appear directly in "Rams," I think that Derrida's theoretical reflection on Celan's poetry essentially substitutes for Levinas's philosophy, whereas Gadamer serves as a substitute for Heidegger early in the essay, only to give way to Heidegger himself at the end.

So the death of Gadamer is an interruption of an ongoing (or "uninterrupted") dialogue between Gadamer and Derrida, and it substitutes for a number of other interrupted encounters, including Derrida's and Celan's, and Celan's and Heidegger's, as well as foreshadows Derrida's own death. Derrida uses Celan's poetry to think through these fundamental issues of encounter, dialogue, hermeneutics, interruption, ethics, ontology, aesthetics, and poetics.

Derrida views death, which we have seen as the ultimate sovereign in the previous chapter, as not simply the interruption of an uninterrupted dialogue, but as catastrophic: "death is nothing less than the end of *the* world."[19] The survivor, in this case Derrida in the wake of Gadamer's death, remains alone "in the world outside the world and deprived of the world."[20] Derrida says that every friendship or dialogue is haunted by the anticipation of the certainty that one friend will die before the other, bringing the dialogue to an end. At the same time, even though it comes to an end, the dialogue in some way continues, in the one who is still alive, beyond the death of the other. Death haunts every friendship and interrupts every dialogue, and yet this very mortality makes dialogue and friendship possible.

After a few pages reflecting on the death of Gadamer, as well as death and dialogue in more general terms, Derrida interrupts his discussion to introduce Celan's poem "Vast, Glowing Vault." He says that like Gadamer, he shares an appreciation and a friendship with Celan, even though Celan is (also) dead. First Derrida quotes the final line, and then he reproduces the entire poem. The last line reads: "The World is gone, I must carry you."[21]

One of the reasons that Derrida wants to read this poem is to repeat Gadamer's hermeneutical efforts to read Celan's poetry, and here Derrida explains how Gadamer understands another poem by Celan, "Paths in the

Shadow Rock." In Gadamer's reading of Celan's poetry, which takes place primarily in the short book *Who Am I and Who are You?*, Gadamer declares that the final line seems to encapsulate the meaning of many of Celan's enigmatic poems. Gadamer does not analyze the poem "Vast, Glowing Vault," but he does affirm that Celan's poetry sets up a dialogical relationship between an I and a You, a speaker and a recipient, that occurs in the poem and does not require any external knowledge on the part of the reader to participate in this encounter. Gadamer states that "I fully agree with the poet that everything is found in the text, and that all biographical-occasional moments belong to the private sphere."[22] Gadamer resists the idea that Celan's poetry requires an esoteric or inside knowledge in order to be read and understood, and primarily contests the interpretations of Otto Pöggeler to the contrary.

One of the poems that Gadamer analyzes in *Who Am I and Who Are You?* is "Paths in the Shadow Rock," and Derrida follows this interpretation, opening it up and attending to the interruptions that Gadamer sees in the poem. The key phase is "Out of the four-finger-furrow / I grub for myself the / petrified blessing." Derrida shows how Gadamer appreciates the fact that the blessing is withheld, it is petrified, and the speaker, the I, grubs for it out of the closed hand, which is the hand of God. Gadamer responds to the withdrawal or petrification of the blessing, but he also subverts and reverses this reading. According to Gadamer, "the beneficial hand is inverted boldly into the hand where palm-reading can reveal a message of beneficent hope."[23]

Despite Gadamer's attempt to read Celan's poem as a message of beneficent hope, he concludes his interpretation with questions about this blessing, including whether it is in fact a blessing. Derrida, who is suspicious of the reversal, says that he admires the respect Gadamer shows for the "indecision" of these final questions.[24] Derrida affirms, both with and against Gadamer's reading, the unreadability and untranslatability of the poem, which also refers to an unreadability of the world: "Even where the poem names unreadability, its own unreadability, it also declares the unreadability of the world."[25] The poem constitutes an abandoned trace of the unreadability of the world to which it attests. Gadamer makes the poem more readable because he has faith that the world is readable in light of an ongoing, potentially infinite hermeneutical process.

Derrida affirms his appreciation of Gadamer, but his affirmation of Gadamer's philosophy is also ambivalent because, for Derrida, every poem harbors "an irreducible remainder or excess" that "escapes any gathering in a hermeneutic." Rather, "the hermeneutic is made necessary, and also possible, by the excess."[26] Over the next few pages of "Rams," Derrida offers his own interpretation of Celan's poetry by reading "Vast, Glowing Vault." He subjects the poem first to a kind of formal analysis and then opens up onto a broader perspective on the issues raised by the poem.

"Vast, Glowing Vault" is an incredibly rich poem, as all of Celan's poems are, and Derrida helps us read this poem in "Rams." The first stanza gives us a cosmological and celestial backdrop, which Derrida calls a tableau. Derrida says that "the black, star-spangled swarm carries the poem away in a hurried, hurrying, headlong movement of properly planetary errancy."[27] The stars themselves are black and they swarm away; they scatter apart. Although Derrida does not mention this, we could also think of the scientific discovery in 1998 of dark energy, an unknown force that is accelerating the expansion of the universe and may eventually pull all matter apart.

The stars have an astrological reference in the poem and invoke the zodiac, especially once the ram is introduced in the second stanza. Derrida claims that the planetary movement is also an animal movement precisely because a ram "will soon bound into the poem: sacrificial animal, battering ram, the bellicose ram whose rush breaks down the doors or breaks through the high walls of fortified castles."[28] The second stanza is the longest and most complex of the poem. "Onto a ram's silicified forehead / I brand this image, between / the horns," an image sealed between the coiled horns of the ram. Derrida suggests that this image branded onto the ram's silicified (or petrified) forehead could be the poem itself. And its resonances are multiple. For one thing, the ram's horn in the Jewish tradition becomes the *shofar*, "the instrument with which music prolongs breath and carries voice."[29] Furthermore, the ram recalls the scene in the Bible where Abraham goes to sacrifice Isaac and his hand is stayed at the last moment. Isaac survives, and a ram is substituted and sacrificed in his place.

This scene is important to Derrida, and he published a book called *The Gift of Death* that was in part a reading and commentary on Kierkegaard's *Fear and Trembling*, which considers this enormous sacrificial event. Here is a longer passage by Derrida from "Rams" that evokes this famous scene:

> Between the most animalistic life, which is named more than once, and the death or mourning that haunts the last line . . . , the ram, its horns and the burning, recall and revive, no doubt, the moment of a sacrificial scene in the landscape of the Old Testament. More than one holocaust. Substitution of the ram. Burning. The binding of Isaac (Genesis 22). After having said a second time, "Here I am," when the angel sent by God suspends the knife Abraham had raised to slit Isaac's throat, Abraham turns around and sees a ram caught by its

horns in a bush. He offers it as a holocaust in the place of his son. God then promises to bless him and multiply his seed like the stars of heaven, perhaps also like the stars of the first stanza. They can also become, in the poem, terrible yellow stars. And it is again a ram, in addition to a young bull, that God, speaking to Moses after the death of Aaron's two sons, commands Aaron to offer as a holocaust in the course of a grand scene of atonement for the impurities, infamies, and sins of Israel (Leviticus 16).[30]

Derrida draws out and makes explicit some of the resonances of Celan's stanza, dealing with sacrifice, substitution, atonement, and holocaust, above all that of the Jewish people as a whole.

The third stanza of the poem asks a question: "In- / to what / does he not charge?" The charge refers to the charge of the ram, that hurls itself about, charging into anything and everything, "in all directions, as if blinded by pain."[31] The sacrificial ram resists the very logic of sacrifice, or sacrificial atonement, just as Celan resists the sacrificial interpretation implicit in the name Shoah. The charge is both a charging and a ramming, but it is also an accusation, as in the charging of someone with a crime. According to Derrida, the charge of the ram suggests "the violent rebellion of all scapegoats, all substitutes."[32] The lamb is the meek animal who goes willingly to meet its sacrificial death, whereas the ram thrashes about, at least in the poem. Derrida imagines that Celan's ram indicts the entire world, because there is nowhere or nothing that he does not charge into: "no one in the world is innocent, not even the world itself."[33] The ethical force of Celan's poetry calls into question the entire world.

Here at this moment in Derrida's reading of Celan's poem I want to pause, to interrupt and to insert another ram from another context and another language. In Chinese, the word for beauty or beautiful is transliterated as měi, and the word měi (美) is composed of two images, one for large (大) and one for ram (羊). According to most scholars, beginning with the late Han lexicographer Xu Shen (ca.55–ca.149), the etymology is culinary because a large ram is not only beautiful but delicious. Xu writes in his dictionary: "when a ram is large, it is beautiful."[34] And so the aesthetic as well as gustatory feeling of the feast provides the origin of the word beauty in Chinese.

Of course, we should have learned from reading Heidegger not to fully trust etymologies, and from Derrida to be suspicious of any claims to origins. In this context, it is interesting that a contemporary Chinese philosopher, Li Zehou, suggests an alternative meaning for the term. In his book on *The Chinese Aesthetic Tradition*, Li argues instead for a ritual meaning of měi, where a large ram headdress sits on top of a person, and the ram is a kind of totem. Here beauty coincides with what is good in a ritualistic context. "The character for 'beauty,' then, with the man on the bottom and the ram on the top, is the manifestation in the written language of this type of animal role or shamanistic totem," according to Li.[35] The point is not to choose between these two proposed origins or etymologies, but to think about their shared resonance with each other, and with Celan's ram. In ancient China, as in ancient Israel, the ram has both a ritual and a culinary function that is connected to sacrifice. Into what culture does he not charge?[36]

There is then an interruption of the ram, but also in the poem an interruption that takes place after the charge of the ram, the ram that charges into and charges against the world. This interruption is a pause, an *Atemwende*, where we catch our breath before the sentence of the last line. The last line stands alone. "It stands," Derrida states, "it supports itself, it carries itself all alone, on a line between two abysses."[37] Now the world is no longer here, it is gone: "Die Welt ist fort." Derrida explains that the world recedes and, in effect, goes away precisely because there is an ethical obligation. He says that "as soon as I am obliged, from the instant when I am obliged *to you*, when I *owe*, when I owe it *to you*, owe it *to myself* to carry *you*, as soon as I speak to you and am responsible to you, or before you, there can no longer, essentially, be any world."[38] The world is the mediating ground of the ethical encounter between you and I, but in the singularity of the encounter the world is gone. We are alone in our encounter, you and I, absolutely alone without world. And yet, there is an obligation. I must carry you, which is an unbearable obligation.

Derrida breaks off his discussion and summarizes five concluding points, dealing with the some of the resonances of the German words *tragen*, which "also refers to the experience of carrying a child prior to its birth," and *Welt*, or world.[39] I only want to focus on the last few paragraphs, where Derrida invokes Heidegger's three theses from his 1929–1930 lecture course, which was published as *The Fundamental Concepts of Metaphysics*. According to Heidegger, the fundamental concepts are: world, finitude, and solitude. In order to elucidate the first question, the question of world, Heidegger introduces three guiding theses: "the stone is worldless, the animal is poor in world, man is world-forming."[40] Understanding the sense in which a non-living rock is without world and an animal has a world but only in an impoverished and tightly limited sense whereas humans are capable of constructing worlds, sheds light on the fundamental question "What is a world?" for Heidegger.

According to Derrida, Heidegger is above all a thinker of the concept of world, and Derrida struggles powerfully in his later years with and against Heidegger's three theses, in works such as *Of Spirit* and *The Animal That I Therefore Am*. For Derrida, there is problematic invidious difference to this unfolding, this dispersal of being as *weltlos* to nonliving stone, *weltarm* to living animal, and *weltbildend* to human being. Each of these beings has its ownmost possibility of being. However, the worlding of the world takes place providentially in and for *Dasein*, the being who has language and can ask the question of being.

One of major themes of Derrida's later work, most explicitly in *The Animal That I Therefore Am*, is the exploration and contestation of this poverty that is attributed to the non-human animal, which is something not only Heidegger but the entire Western philosophical tradition presupposes, including Levinas and in some respects even Celan.[41] Drawing on Jacob von Uexküll's influential studies of animal behavior, Heidegger claims that the animal is poor in world because it is captivated by its instinctual drives in "an intrinsic self-encirclement."[42] This encirclement is tightly drawn around the animal like a bee or a tick such that it opens up a very limited sphere of behavior beyond with the animal cannot experience. According to Heidegger, "the life of the animal is precisely the struggle to maintain this encircling ring or sphere within which a quite specifically articulated manifold of dispositions can arise."[43] The animal has no awareness of a world beyond this encircling ring.

On the one hand, Derrida points out that Heidegger claims that the animal does not die because it does not have a relation to being that takes the form of an "as such."[44] It cannot die because it cannot be related to death as such as its ownmost possibility of being. But on the other hand, as Derrida explains, Heidegger admits that what distinguishes the animal from the stone is the fact that it can die because it is a living being. It possesses "the living character of a living being," which is fundamentally related to the possibility of dying.[45] So the animal sort of lives and sort of does not fully live. According to Derrida, Heidegger fails to confront or fully engage with the animality of Dasein as a living being because "Dasein is explicitly described by Heidegger as a being that is not, essentially, a 'living' being."[46]

Derrida wants to undermine the identification of the essence of *Dasein* as a world-builder, one who creates worlds with language and participates in the worlding of the world that being (*Sein* or *Seyn*) accomplishes. According to Heidegger, the "poverty in world" that characterizes the animal "implies a deprivation of world."[47] But in Celan's poetry, Derrida finds a provocative deprivation of world for human being. Here Celan's poetry, and in particular this poem with its powerful final line, interrupts and breaks this secure process of world formation at work in the implicit progression from stone to animal to human. Derrida claims at the end of "Rams" that "for reasons I cannot develop here, nothing appears to me more problematic than these three theses."[48] For reasons he elaborates on more explicitly in *The Animal That I Therefore Am*, therefore, nothing appears more problematic than Heidegger's theses about stone, animal, and world.

Derrida asks about the being-gone (*Fort-sein*) of the world in Celan's poem, and suggests that it proceeds according to a completely different logic than Heidegger's. If the world is gone here, for human beings, at the heart of the asymmetrical ethical relationship with the other, then "isn't it the very thought of the world that we would have to rethink, from this fort, and this fort itself from the '*ich muß dich tragen*'"?[49] This, Derrida says, is one of the questions that he would like to pose to Gadamer, "appealing to him for help," "in the course of an interminable conversation." This conversation, of course, has been interrupted by death, but it is always already interrupted by death, and this death and this interruption are what make conversation and dialogue possible in the first place. Gadamer's death is the end of the world, and so is Derrida's, and so is Celan's. We who live bear witness to the end of the world.

This is Derrida's faith, although it is not a conventional faith. As I have already considered in various ways, in "Faith and Knowledge," Derrida claims that what we call religion always has two sources, one of which is the sacred or holy, and the other is faith or the experience of belief. These two sources conflict and contaminate each other, preventing any certain assurance or indemnification. As Michael Naas explains in *Miracle and Machine*, which I have also discussed, Derrida prefers the second source of religion, faith, as an experience of social trust or belief, over the source of the sacred as the unscathed. At the same time, Naas points out that "Derrida is quite clear about the disruptive nature of this faith."[50] Derrida's faith, like that of Celan, is disruptive rather than constructive; belief undermines the machinic nature of religion even as it functions as an expression of it. These two sources can be distinguished but not fully separated, and they work in nature and in natures in the world, not simply in language.

Being and world, like the poem, exist only insofar as they are in a state of being gone, a gone-ness that as interruption gives the possibility for ethical relation. A poem is also a kind of machine, infected and inflected with an originary technicity. This technicity exceeds any simple mechanical understanding of a machine because it incorporates both the machinic repetition and the exceptional miracle that exceeds it.

Worldlessness is a kind of original possibility for a world to be, to exist. The world is not absolutely disappeared for Derrida, but it withdraws in order to foreground the ethical relation, ungrounded from any literal writing. In a disruptive way that attests to the worldlessness of world, Celan's silence, cryptic poetry, and even suicide offer interruptive breaks that are the basis for ethical relations today. Derrida profoundly appreciates the ethical force of Levinas's *Otherwise than Being*, and this is the same Levinas who asks of Celan: "does not he suggest a modality *otherwise than being*?"[51] According to Levinas, Celan's poetry represents an "insomnia in the bed of being, the impossibility of curling up and forgetting oneself. Expulsion out of the *worldliness of the world* . . ."[52]

Despite his affinity for Levinas, however, Derrida wants to hold on to just a thread of being, a threading of worlds, like the thread-suns or *Fadensonnen* of Celan: "there are/still songs to be sung on the other side/of mankind."[53] But there are no more songs for us when the world is gone. Every death is the end of the world, which is different and unique every time. And in the absence or withdrawal of world, which hangs only by a thread, I must carry, I must bear (*tragen*) you, even if I cannot bear you. There is no other way. Every world is an island separated by an unshareable abyss that constitutes an almost unimaginable archipelago, as Derrida suggests in the second volume of *The Beast and the Sovereign*.[54] This conclusion is just one of the rich threads of Celan's disturbing and powerful work that Derrida brings to bear. And in the next chapter, I will further reflect on the theme of the end of the world in Derrida as it relate to the object-oriented ontology of Timothy Morton and other speculative realists.

But first, wait, I want to pause and insert into this potentially infinite conversation another ram from the Bible. This one is not from the *Aqedah*, but from the book of Daniel, chapter 8, verses 3–7:

> I raised my eyes and there I saw a ram with two horns standing between me and the stream. The two horns were long, the one longer than the other, growing up behind. I watched the ram butting west and north and south. No beasts could stand before it, no one could rescue from its power. It did what it liked, making a display of its strength. While I pondered this, suddenly a he-goat came from the west skimming over the whole earth without touching the ground; it had a prominent horn before its eyes. It approached the two-horned ram which I had seen standing between me and the stream and rushed at it with impetuous force. I saw it advance on the ram, working itself into a fury against it, then strike the ram and break its two horns; the ram had no strength to resist. The he-goat flung it to the ground and trampled on it, and there was no one to save the ram.

Is this ram in Daniel similar to the ram that Abraham substituted for Isaac in Genesis? What if the ram is not just a sacrificial animal but, in fact, a name of God? According to Jacques Lacan, the ram is an Elohim, as well as a Name-of-the-Father. For Lacan, and I suggest also for Derrida, our discourses about animality, especially in its sacrificial element, are caught up in our theological understandings about divinity.

In November 1963, two days prior to John F. Kennedy's brutal assassination in Dallas, Lacan delivered his final seminar at the Hôpital Sainte-Anne. This seminar was to focus during 1963–64 on "The Names-of-the-Father," which was announced at the end of his previous seminar, Seminar X, on Anxiety. Unfortunately, the political expulsion of Lacan as a training analyst by the French Psychoanalytic Society terminated his Seminar at Saint-Anne just before it began, and Lacan only gave one lecture, an "Introduction to the Names-of-the-Father" on November 20, 1963. His seminar was interrupted, and it would reconvene later at the École Normale Supérieure, at the invitation of Louis Althusser, in early 1964, but Lacan never returned to or completed his seminar on "The Names-of-the-Father." Instead he taught his famous seminar XI on "The Four Fundamental Concepts of Psychoanalysis."

The lecture from November 20, 1963 was not published until 2005 by Jacques-Alain Miller. In it, Lacan reflects on the scene of the *Aqedah*, as expressed by Kierkegaard's *Fear and Trembling*, Caravaggio's paintings of *The Sacrifice of Isaac*, and the medieval French rabbinical commentator Rashi. Caravaggio paints two distinct canvases of this primary scene, featuring Abraham, Isaac, an angel, and a ram. Lacan points out that in one of these paintings, "there is one in which the ram is on the right and where you see the head that I introduced last year, invisibly, in the form of the shofar—the ram's horn. This horn has indisputably been ripped off of him."[55]

In Seminar X on Anxiety from the year before, Lacan discusses the shofar as an instantiation of the *objet petit a*, the object cause of desire. The object *a* encapsulates desire and focuses it on an object, but in such a way that the object expresses the infinite nature of desire. The object *a* substitutes for any and all objects of desire in a metonymic way. The first object *a* for Lacan is the breast because the breast represents all of human nourishment, care, vitality, and sexuality for the infant.

In Seminar X, Lacan says that there are five stages of the object *a*, and the shofar expresses the fifth stage, the object of desire at the level of the ear, the voice or sound. The shofar encapsulates and represents the voice of Yahweh for the Israelites. Lacan relates that "the horn is generally, though not always, a ram's horn."[56] The shofar is an object that gives voice to the voice of God. The object *a* expresses an element of the Real because it represents something that exceeds the symbolic order. And every god, including Yahweh, is "an element of the real, whether we like it or not, even if we no longer have anything more to do with them."[57]

According to Lacan, any time one encounters a God, "a God is encountered in the real."[58] The God of the Israelites is made manifest in and through the shofar, which produces the sound or voice of God. Lacan says that according to Rashi, the Jewish commentator, "the ram in question [at the sacrificial scene of Abraham and Issac] is the primal Ram." This ram "was there, writes Rashi, right from the seven days of creation, which designates the ram as what he is: an *Elohim*."[59] If the ram is an Elohim, a God, then the sacrifice of the ram is more than a simple animal sacrifice.

The ram does not simply substitute for the human son, Isaac; he substitutes for Yahweh as another form of God. The ram rushes onto the scene of sacrifice on Mount Moriah and is caught in a thicket.[60] Lacan claims that "the one whose Name is unpronounceable [Yahweh] designates him to be sacrificed by Abraham in his son's stead. This ram is his eponymous ancestor, the God of his line."[61] The primal Ram is sent to be a sacrifice, who justifies the sacrificial death of all other rams. The ram is the sacrificial animal par excellence, which is why it is an Elohim and a name of God.

Lacan argues that this situation, this charging of the ram onto the sacrificial scene, shows "the sharp divide between God's jouissance and what, in this tradition, is presented as His desire."[62] Ordinary desire is what we want, or what God wants in a simple or direct way. Jouissance, for Lacan, represents desire taken to its limit, beyond any reasonable bounds. Jouissance manifests a death drive because it operates beyond the pleasure principle for the subject. Here Lacan, in psychoanalytic terms, considers how humans think about and represent the desire of God, which always threatens to go crazy, beyond all limits.

This gap between desire and jouissance at the level of God is filled by the ram, a symbolic substitute whose meaning is more than symbolic, and whose death allows the Israelites to affirm their lineage to Abraham despite the separation indicated by Isaac's symbolic death. Israelite genealogy is wholly symbolic, not real, because this (non-)sacrifice accomplishes the task of diminishing "the importance of biological origin."[63] This diminishment of biological origin is especially relevant insofar as the figure of Isaac serves to stitch the cycle of stories associated with Abraham to the cycle of stories associated with Jacob. Many biblical scholars, in fact, question the historical existence of Isaac.

In his lecture "On the Names-of-the-Father," Lacan complicates the relationship among humanity, animality, and divinity. In his work, Derrida pursues the limits of the human being as it exists in a state of tension between animality and divinity. The human is connected at the animal level to the beast, and at the divine level to God as Sovereign, and incorporates both elements in a discordant accord. Each of these figures of animal, human, and God is caught in the tension between the miraculous singularity of life and the deathly repetition of the machine.

Here we can see how the figure of the ram from Celan's poem as well as the ram in the story of the binding of Isaac, not to mention the desolated ram mentioned in Daniel, animate these questions about the nature and scope of Heidegger's philosophy, including his three theses about the stone, the animal, and the human being. Much of Derrida's later work focuses on the animal, contesting Heidegger's claim that the animal is poor in world, and showing how the theoretical discourse on the fundamental distinction between the animal and the human from Descartes through Levinas, including that of Lacan, deconstructs. We cannot simply maintain a strict boundary between the human animal and the nonhuman animal.[64]

In the next chapter, I consider further the situation of the inanimate object, in Heidegger's case the stone. Although Derrida does not pursue this line of thinking, we can think about how his questioning of Heidegger's threefold typology implies a challenge to the claim that the stone is simply without world in the way that animals possess a world poorly, and that humans compose worlds for themselves with the help of being. In Volume II of *The Beast and the Sovereign*, Derrida points out that Heidegger offers the stone as a particular example of an inanimate object, while for the other two kinds of things, he "says in a general way, with no examples, 'the animal' and 'man.'"[65] Derrida takes this example of a stone as a kind of stumbling block for Heidegger's thinking but then fails to pursue it further in connection to a stone or another inanimate object. This line of thinking, however, brings Derrida up against a newer philosophy that usually is seen as incompatible with Derrida's thought, an object-oriented ontology, as we will see in the next chapter in the context of contemporary philosophy of religion.

Derrida, Lacan, and Object-Oriented Ontology
Philosophy of Religion at the End of the World

There is a world within the world.

—Don Delillo, *Libra*

Jacques Derrida (along with Levinas) is the major point of reference for the formation of a Continental philosophy of religion that took shape in the 1990s around the work of John D. Caputo, Richard Kearney, Edith Wyschogrod, Kevin Hart, and Merold Westphal. Unfortunately, much of what passes for philosophy of religion today, whether analytic or continental, is disconnected from the pressing realities of lived existence, caught up in conceptual arguments and juridical formulations about non-existing entities. Why should we care about philosophy of religion, much less invest in the effort to rethink it?[1] In this chapter, I reflect briefly on the history and present situation of philosophy of religion, and then suggest an ecological becoming of philosophy of religion in both generic and technical terms, in relation to the thought of Derrida and Lacan.

This thinking about philosophy of religion takes place at the edge of the world, perhaps on the precipice of a biological catastrophe that Richard Leakey, Elizabeth Kolbert, and others call "the sixth extinction."[2] All of our thinking in this century has to grapple with the urgency of our ecological situation, which includes resource depletion, global climate change, over-population, ocean warming and acidification, and mass extinction of many species of birds and mammals. I will return to this catastrophic horizon at the end of the chapter, by way of an engagement with Timothy Morton's ecological thinking in *Hyperobjects*.

In this chapter, I engage Derrida's philosophy with the newer viewpoint of "object-oriented ontology" (OOO) the idea that we need to shift our attention from subjective to more objective modes of understanding. Morton provides an impressive reading of an object-oriented ontology in *Hyperobjects: Philosophy and Ecology after the End of the World*, and I will engage directly with Morton's book shortly, using Derrida and Lacan to help us think more carefully about what an object is. OOO, a phrase coined by Levi Bryant, is also associated with the object-oriented philosophy of Graham Harmon and a philosophical movement called Speculative Realism that has emerged in the first decade of the twenty-first century.[3]

A powerful influence on Speculative Realism, as well as a strong impetus toward this OOO is supplied by the French philosopher Quentin Meillassoux in his book *After Finitude*. In this influential book, Meillassoux critiques and tries to get out of the subjectivist trap of correlationism, in which any knowledge of an object has to be correlated with a thinking subject. Correlation, he writes, "consists in disqualifying the claim that it is possible to consider the realms of subjectivity and objectivity independently of one another."[4] Speculative Realism wants to effect a turn away from the subjectivism that pervades post-Kantian Continental philosophy, and turn our attention toward the strange and complex behaviors of objects. In so doing, they also bring about a re-engagement with some of the perspectives and observations of the natural sciences and mathematics.

In *Hyperobjects*, Timothy Morton cites Derrida a number of times with approval, but many writers associate Derrida's philosophy with an irreducible subjectivism if not a hopeless linguistic idealism. For Harman and Meillassoux, Derrida's work is caught in the trap of post-Kantian correlationism, where any object necessarily has to be correlated with a thinking subject.[5] Modern correlationism begins with Kant's Copernican Revolution, that claims that objects rotate around human categories of intuition and understanding. This correlationism persists all the way through most Continental philosophies of post-structuralism, including Derrida's philosophy.[6]

As a way to help us think beyond the impasse of correlationism, Meillassoux introduces the notion of the arch-fossil. An arch-fossil names an object that bears traces of "the existence of an ancestral reality" that is "anterior to terrestrial life."[7] An arch-fossil is an object of thought that signifies something that occurs before any human thinking exists. Meillassoux claims that the arch-fossil is a natural artifact that testifies to a limit of human experience, but instead of drawing a conclusion about human finitude, he exploits the notion of an arch-fossil to push beyond the correlationist circle. The way that he does this is to radicalize the contingency inherent in correlationism.

Even if everything is contingent, including any object and any subject, Meillassoux argues for the necessity of contingency because it is the "factiality" or contingency of the correlation itself that is absolute, rather than any particular entity.[8] The arch-fossil exhibits an absolutization of facticity that Meillassoux calls factiality, because it indicates a kind of givenness of being that is anterior to any thinking of givenness. Meillassoux discovers a method of radicalizing correlationism itself to escape the correlationist circle because he claims that we cannot presuppose that any necessary being exists, but we can in fact claim that it is necessary and absolute to hold that any being might not exist. Every being is contingent: "the absolute is the absolute impossibility of a necessary being."[9]

From this principle of unreason, or refusal of any principle of sufficient reason, Meillassoux addresses the appearance of stable laws in our world and how they can exist in a universe of radical contingency. Taking his cue from David Hume, Meillassoux argues that despite all scientific laws being contingent, nevertheless their chaotic flux constructs an apparent stability. Meillassoux calls this "Hume's Problem," which concerns how our understanding of causality presupposes the uniformity of the laws of nature. In Part VIII of his *Dialogues Concerning Natural Religion*, Hume speculates about the possibility that a finite number of particles undergoing a finite number of transpositions over an indefinite period of time would generate an eternal repetition of the same situations. This scenario foreshadows Nietzsche's articulation of the idea of eternal recurrence. If matter is put into ceaseless motion by its originating force, then this ongoing dynamism produces "a continued succession of chaos and disorder." But Hume has his character Philo suggest that in this very situation, "is it not possible that it may settle at last, so as not to lose its motion and force . . . yet so as to preserve a uniformity of appearance, amidst the continual motion and fluctuation of the parts?"[10]

Meillassoux adopts a similar solution in *After Finitude*, but he presents it in mathematical set-theoretical terms, an area where he is influenced by Badiou's philosophy in *Being and Event*. Meillassoux claims that Georg Cantor's notion of a transfinite set of numbers (basically multiple sets of infinite numbers) demonstrates that "the (quantifiable) totality of the thinkable is unthinkable."[11] Cantor's set theory mathematics allows one to compare infinite sets, and to say that one infinite set is larger than another, even if both are infinite. But this series of infinite sets is unthinkable in a total sense; the series of transfinite cardinal numbers cannot be totalized because they are based on sets whose members are infinite. For Meillassoux, the possibility of a non-totalizing (in a specific quantifiable sense) possibility as exhibited in transfinite sets (sets of infinite sets) allows him to posit a certain stability of experience without the necessity of underlying laws. This is a complex technical argument, but Meillassoux is relying on

Badiou's work to claim that reality is essentially formalizable in mathematical terms with set theory, and Cantor's work suggests that we cannot totalize all possibilities, which means that there is a sense in which our laws are contingent but nonetheless stable. Meillassoux claims that all of our laws of reason and nature are contingent and potentially chaotic, but "it is precisely this super-immensity of the chaotic virtual that allows the impeccable stability of the visible world."[12]

Much of the impetus behind Meillassoux's critique of correlationism concerns how it sustains postmodern views of religion. For Meillassoux, correlationism ends up with an irresolvable aporia between what we can know about ourselves and what we can know about the world. This leaves open the door for fideism, famously affirmed by Kant in the Preface to the Second Edition of the *Critique of Pure Reason*. Kant says that "I have therefore found it necessary to deny *knowledge*, in order to make room for *faith*."[13] This faith is the real target of Meillassoux's critique of correlationism because in correlationism the idea of the absolute is "reduced to a mere belief, and hence to a religion, albeit of the nihilist kind."[14] Correlationism reaches an impasse of knowledge because it equivocates between our objective knowledge and its necessity to be correlated with the subjective processes of our knowing. We cannot know anything about the objective or real world, so we are forced to the state of mere belief. Because we lack any objective vantage point to think the absolute, philosophy reverts to a form of weak piety. "On this point," Meillassoux concludes, "the contemporary philosopher has completely capitulated to the man of faith."[15]

Meillassoux does not name Derrida here, but his critique is certainly concerned with the turn to religion on the part of many Continental philosophers. Derrida does not reduce knowledge to faith, but there is this aporia or equivocation in Derrida's thought between faith and knowledge that troubles Meillassoux and many other Speculative Realists. However, as Steven Shakespeare points out, "Meillassoux's position comes with a price. Having set out to establish the validity of scientific statements about an ancestral past, we are left with a world of chaos in which there is no reason for anything to happen or not happen. The fossil dissolves into abstract possibility."[16] In his effort to avoid the indeterminacy of subjectivism, Meillassoux cannot evade the paradox of objective contingency. The only way he can escape correlationism is to radicalize contingency, but the radicalization of contingency introduces an objective equivocation that is no less disturbing.

In his opposition to contemporary philosophical fideism and its weak form of religion, Meillassoux introduces a strange return of God. In *The Divine Inexistence*, Meillassoux claims that even though he is a strict atheist, and God does not exist in any actual or potential sense in this world, it is conceivable that God could exist in virtual terms if a new world came into existence that provided the conditions for the advent of justice. Meillassoux claims that we have experienced three fundamental advents of something entirely new, or "three *orders* that mark the essential ruptures of becoming: *matter, life, and thought.*"[17] Human value and morality desires another new order, one of justice, but that order does not exist, and cannot exist without the restoration and resurrection of humans who have already died. Even though this just situation is potentially impossible in any way in our own world with its current laws, Meillassoux can imagine a virtual world where such a rupture could occur that would bring about a world in which justice exists, and Meillassoux suggests that we have a messianic imperative to think and proclaim such a virtual reality.

To claim that God exists is blasphemous because it affirms our present order of injustice, but atheism goes too far in its giving up of any possibility of God. Meillassoux uses the word God to name the possibility (or virtuality) of a world in which justice is possible in a strong sense, which includes some kind of resurrection of the dead and involves a Christ-like human mediator who abandons power for the sake of justice. Meillassoux claims that authentic philosophical faith, in contrast to religious faith, consists of "*believing in God because he does not exist*," and this is an option that has not yet been seriously considered.[18]

Meillassoux's conclusion about the inexistence of God appears incredible and bizarre in ontological terms, even if it is a provocative speculation. Why this God specifically, and why should it conform so closely with more conventional monotheistic and even Christian notions of God? Meillassoux's God is very much a God of the philosophers, and stems from a very European Enlightenment approach to the situation of human moral existence. We may and should resonate with his deeply felt moral desire for justice, possibly even to the point of a messianic hope for an entirely new advent. Meillassoux bases his messianic hope on an affirmation of the ultimate value of the human. This humanism seems naïve, however, not only from the perspective of the deconstruction of humanism of poststructuralist Continental philosophy, but also from the ontological and cosmological perspective of the universe that emerges from post-humanist versions of technology and natural science. Speculative realism and speculative materialism has many invigorating aspects, but some of these conclusions appear escapist and suffused with a kind of wishful thinking, despite the brilliance of Meillassoux's thinking and writing.

In our postmodern as in our modern world, then, we seem to not be able to have done with the question of religion, despite our best efforts to be atheists. Genuine atheism is difficult, as Christopher Watkin insists, although perhaps we need to stop trying quite so hard to eliminate religion.[19] In the modern world, philosophy of religion stems from Kant and Hegel, which makes it intrinsically correlationist in terms of Speculative Realism. In a negative sense, Kant determines the limits of a philosophical understanding of religion because he excludes it as a legitimate object of critique. Reason cannot critique religion in its pure or practical role because religion does not occupy an autonomous realm of knowledge. For Kant, a purified rational religion is a supplement to ethics, which is the realm of practical reason that corresponds to the question: "What should we do?"

In a Kantian sense, there persists a tension intrinsic to philosophy of religion because philosophy is the rational and critical function that defines the state and stakes of the situation, while religion indicates an object that in some ways resists rational and critical explanation. Philosophy of religion tries to explain what cannot be entirely explained by philosophy. Hegel, on the other hand, gives philosophy of religion a positive function because he argues in his *Lectures on the Philosophy of Religion* that religion possesses content in the form of *Vorstellungen*, or images, that can then be spelled out in conceptual forms according to the dynamic development of the Concept, or *Begriff*.[20] For Hegel, philosophy of religion is the penultimate stage of the outworking of absolute Spirit, the point where Spirit as Subject recognizes its intrinsic form in images and then elaborates their conceptual content.

For Hegel, religion has importance, but it is only penultimate significance, and philosophy must supply its conceptual content for Spirit to become all in all. Philosophy of religion is a spiritual process that evolves the conceptual Truth that emerges from religious forms. Stereotyping very broadly, we could suggest that Anglo-American analytic philosophies of religion conform more closely to the Kantian model of religion, whether they want to redeem or repudiate religion in critical terms. In a complementary way, most Continental philosophies of religion can be viewed as Hegelian because they are more interested in elaborating religious ideas and then re-describing them in other ways, even if they are seen as post-Kantian in terms of their correlation to subjective modes of knowing. Sometimes this redescription is affirmative of religious phenomena and sometimes it is dismissive, but it is the process of conceptual elaboration that is important.

Whether seen as more Hegelian or more Kantian, Continental or Analytic, philosophy of religion is fairly specialized as a subset of academic scholarship that is increasingly marginalized in our contemporary corporate university. Professional philosophy itself is so alienated from popular discourse that it is difficult to translate these ideas into language that can be understood by, let alone evaluated by, a broader public. At the same time, philosophy more generally struggles to justify itself before an academic tribunal that is engorged by administrative functions, and demands disciplinary currency that can be calculated ultimately in quantitative, monetary terms. These problems affect contemporary philosophy and philosophy of religion whether or not they are criticized as intrinsically correlationist.

For scholars of religion, religion is an object of academic study that is at least partly "the creation of the scholar's study," as Jonathan Z. Smith affirms.[21] I disagree with Smith, however, when he claims that religion is "*solely*" the creation of the scholar's study because I think that goes too far in its academicism. The philosophical viewpoints of OOO and Speculative Realism assert that such a position reflects the inherent subjectivism in the modern and contemporary academy, which they are attempting to dislodge. There are objects of and for religion. But what sort of objects are they? I will consider this question after briefly considering how religion is seen as contributing to a worldview and the constitution of a world.

Scholars like to explain a word's meaning by appealing to etymology, and the word "religion" can be traced back to the Latin *religio*. *Religio* has two competing etymologies, and scholars have been unable to finally decide on which one is correct. The most popular meaning is *re-ligare*, meaning to re-bind or bind back. This suggests that religion is a form of re-binding of a social fabric that has been torn or broken. The problem with this understanding is that it presupposes an original harmony or unity that comes undone, after which the role of religion is to put it back together.

The other, competing, etymology of *religio* is *re-legere*, which means to re-collect or re-read. I like this meaning because it suggests a kind of repetition, and this repetition does not necessarily mean that what is re-read is the same text or the same practice. Furthermore, we could reread the first meaning of *religio* as *re-ligare* in terms of what Caputo calls "a binding to the unbound," where life is unbound from anything other than or transcendent to life. I will return to consider Caputo's reading of Derrida on religion more explicitly in the next chapter. Here, if this binding back is to what is originally bound, then it suggests an originary purpose or meaning to existence. But if religion is binding to the unbound, this frees up religion for other meanings, practices, and bindings.

Whatever religion as an object of study is, it has famously returned over the last few decades, and this return has falsified or at least severely problematized the so-called "secularization hypothesis." The secularization hypothesis suggests that religion is becoming less significant for human society and meaningful practices, and its role is being replaced by other things. The return of religion in philosophy, politics, and culture attests to a limit of this hypothesis.

Of course, religion does not simply return because it never went away. What happens, as sociologists such as José Casanova explain, is that religion has become deprivatized.[22] During the period of European modernity, religion was seen as a private matter of belief, in contrast to public secular reason. Religion could not be kept completely private, but the idea that it could be informs the ideology of secularism and fuels the secularist hypothesis. I argue that the return of religion indicates a postsecularism, where the simple opposition between religious and secular deconstructs. At the same time, I want to resist the postsecular narrative advocated by Radical Orthodoxy and other religious apologists, which sees religion as *replacing* the secular in a triumphalist way. I prefer the more nuanced approach of Talal Asad, who argues that religion and politics are both implicated in the constitution or formation of the secular. Asad concludes that "if the secularization thesis no longer carries the conviction it once did, this is because the categories of 'politics' and 'religion' turn out to implicate each other more profoundly than we thought, a discovery that has accompanied our growing understanding of the powers of the modern nation-state."[23]

Derrida has been one of the primary theoreticians of this intertwining of religion and politics, as we saw in Chapter 3. His analysis of religion in "Faith and Knowledge" demonstrates a kind of auto-immunity at work in the political return of religion, because adherents of political forms of religion desire to ward off attacks to the social and political body, but these efforts to protect and indemnify society end up harming it. In *Rogues*, Derrida says that there exists a "perverse and autoimmune" effect not only of religion, but even of democracy, whereby democratic states must "interrupt a normal electoral process in order to save a democracy threatened by the sworn enemies of democracy."[24] This occurs in Algeria in 1991 when the government intervened to halt an election that gave Islamists political power, and in the United States after 9/11, when the United States Patriot Act restricted citizens' rights in the name of protecting them from future terrorist attacks.

The religious, the political, and the secular are all constituents in the composition of a world. According to Martin Heidegger, human beings are world-making beings. In his *Fundamental Problems of Metaphysics*, as we saw in the previous chapter, Heidegger examines three ideas: world, finitude, and solitude. In order to elucidate the first question, the question of world, Heidegger introduces three guiding theses: "the stone is worldless, the animal is poor in world, man is world-forming."[25] For Heidegger, man is *Dasein*, the 'there' of being who is capable of asking the question of being. Stones lack any ability to ask questions, and animals are locked into their worlds, at best only dimly aware of their existence in a world. Leaving aside for the moment the question of whether Heidegger is unfair to animals and rocks, we can see how this conception of world is important for Heidegger.

In his later work, he argues that rather than humans intentionally constructing worlds, it is being itself that allows worlds to be created in and through human language and activity. In his essay on "The Origin of the Work of Art," Heidegger claims that setting up a work of art such as a temple opens up a world. "Towering up within itself, the work opens up a world, and keeps it abidingly in force."[26] A world is not simply what lies at hand, how we understand our perspective on or representation of the world. As disclosed in a genuine work of art, "the *world worlds*, and is more fully in being than the tangible and perceptible realm in which we believe ourselves to be at home."[27] Stones, plants, and animals do not possess a world because they cannot wield language or create a work of art. According to Heidegger, "a peasant woman, on the other hand, has a world because she dwells in the overtness of beings." A human person possesses the capacity to experience a world as a world, in the becoming or worlding of this world, in which "all things gain their lingering and hastening, their remoteness and nearness, their scope and limits."[28] Setting up a work of art means setting forth a world.

To be human or to be able to ask the question of being is to participate in the creation of a world. And we are only able to take part in this world-creation insofar as we are finite beings, aware of our own mortality or what Heidegger calls "being-unto-death." A world is temporary and finite, but it opens up and discloses itself to contemplative humans who let the world emerge into being. I think that part of what it means to be religious is to be oriented within a world, that religion as a work of creation provides a sense of orientation—the scope and limits of all things—even as it also allows for and creates a certain amount of disorientation. This orientation is provided abruptly and crudely in movie previews that invariably begin: "In a World . . .".

A world anticipates its own demise. But what if the end of a world becomes so gigantic and extreme that it calls into question not only its own existence but the existence of any possible world? This disturbing question is what Morton's book *Hyperobjects* forces us to think. What is a hyperobject? It is a special kind of object, and it attests to the popularity as well as the limits of this newer philosophy of OOO. OOO, as we have seen, wants to shift philosophy's focus from language and subjectivity to objects.

Proponents of OOO want to think about objects without their being tied to the conditions of representation given by a human subject, as laid out in Kant's critical philosophy.

Morton embraces this object-oriented-ontology, but he expands the definition of an object to include what he calls hyperobjects. A hyperobject is something that is "massively distributed in time and space relative to humans."[29] As examples of hyperobjects, Morton lists the Solar System, a black hole, an oil field, the Florida Everglades, and the biosphere, but his main example is global warming. For Morton, hyperobjects come into view when we abandon an anthropocentric perspective and adopt a more objective realism in our philosophy. What is ironic is that it is precisely in the era that is being called the Anthropocene where we come to appreciate the extent to which humans have transformed the planet, that we can see this shift in orientation. According to Morton, the transformation of Earth from a natural world to one where we cannot escape our responsibility for transforming nature is tied to the end of the world.

Hyperobjects emerge at the end of the world, and the end of the (natural) world has already occurred. The first instance of the end of the world happens in 1784, with the patenting of the steam engine, "an act that commenced the depositing of carbon in Earth's crust," which marks "the inception of humanity as a geophysical force on a planetary scale."[30] This end of the world is repeated in 1945, with the testing of the first atomic bomb in Trinity, New Mexico. Morton says that

> what comes into view for humans at this moment is precisely the end of the world, brought about by the encroachment of hyperobject, one which is assuredly Earth itself, and its geological cycles demand a *geophilosophy* that doesn't think simply in terms of human events and human significance.[31]

Again, I want to underline the paradoxical nature of Morton's analysis here: It is the recognition of irreducible human effects on the Earth that ushers in the Anthropocene and ends any simple understanding of Nature as a sphere apart from human activity that at the same time creates the possibility of thinking about objects in a non-anthropocentric way. In addition, hyperobjects are very strange sorts of objects, and resemble something that could also be called systems or even processes.

I note Morton's insistence that the emergence of hyperobjects inaugurates a present and a future "after the end of the world," because we can no longer experience a world without human beings being at the center of it. Hyperobjects bring about a quake in being that shakes us out of our sense of who and what we are when confronted with such tremendous entities. Morton focuses mainly on very large hyperobjects in his book, and perhaps he can be accused of gigantomachy in his OOO, his desire to prioritize the very large over the very small. But what's interesting about his analysis is his elaboration of the stakes of a flat ontology, where there is no container or horizon within which objects fit. This is the specific sense in which hyperobjects proclaim the end of the world. The expansion of our ecological awareness taken to its end brings about the realization that we no longer live in a closed environment. This means that the more we become aware of the interconnectedness of things, "the more it becomes impossible to posit some entity existing beyond or behind the interrelated beings."[32] These interrelated beings cannot be fitted into a container or "umbrella that unifies them, such as world, environment, ecosystem, or even, astonishingly, Earth."[33]

Morton argues that hyperobjects bring about the end of the world, because they make it impossible to conceive a world as a container for the set of objects that resides in it. No container, no world. Earth is one object among other objects, and all of these objects exist on the same flat plane of immanence. Hyperobjects are special objects because in their vastness they dwarf what we normally consider objects. But they are still objects for Morton, and so they contribute to an OOO.

I want to back away slightly from the extreme philosophical situation that Morton depicts. Or rather, I want to generalize it. For me, all objects are hyperobjects, which can also be called systems. The problem with objects is that they have no natural simple boundaries. And rather than accept Morton's claim that at some determinate moment or moments in history, we experience a qualitative change that signifies the end of the world, I suggest that we already live at the end of the world. The subtitle of Morton's book is "Philosophy and Ecology After the End of the World," whereas I want to think ecologically about philosophy and religion "at the end of the world" using Derrida as a resource.

According to Derrida, Heidegger's three theses about the world—the stone is without world, the animal is poor in world, and the human being is a world-builder—are extremely suspect. Derrida says that for him, "nothing appears to me to be more problematic than these theses."[34] In his later philosophy, including *The Animal That I Therefore Am*, Derrida takes up the question of the animal to challenge Heidegger's view that animals are poor in world while human beings are builders of worlds. As discussed in the previous chapter, in "Rams," Derrida reflects on the recent death of Hans-Georg Gadamer, and considers the poem by Paul Celan called "Vast, Glowing Vault." Here, the ram is a sacrificial animal that substitutes for Isaac in the famous account of the *aqedah* or binding, as narrated in Genesis

22 and then famously dramatized by Soren Kierkegaard in *Fear and Trembling*. In the poem by Celan, the ram charges, and the question is inverted: Into what doesn't the ram charge? Finally, the last powerful line reads: "The World is gone, I must carry you."

In his reading of Celan, Derrida challenges Heidegger, suggesting that the direct relation between the I and the you occurs precisely when the world is gone, in German *fort. Die Welt ist fort*. The world is gone. Derrida says that "as soon as I am obliged, from the instant when I am obliged to *you*, when I *owe*, which I owe it *to you*, owe *it to myself* to carry *you*, as soon as I speak to you and am responsible for you, or before you, there can no longer, essentially, be any world."[35] The immediate ethical relationship of responsibility to the Other means that the world is gone, it goes away, it becomes *fort*. And this responsibility is not restricted to human beings, as it often appears to be in the philosophy of Emmanuel Levinas. We can be directly and ethically related to an animal, such as a ram.

But what about a stone? Heidegger says that a stone is without world; it simply does not and cannot possess a world. Furthermore, Derrida does not consider stones specifically in his attempt to complicate and refute Heidegger's three theses. But Derrida does say that nothing seems more problematic than these theses, and he does not specify that he only disagrees with the first two. Is the pure lack of world that Heidegger accords a stone similar to the gone-ness of world that Derrida finds in Celan? It would seem not, but at the same time it would seem strange to disallow deconstruction to work on stones if it can work on and with animals. Can anyone be responsible to a stone? Can a stone be an Other in the technical poststructuralist sense? What kind of object would a stone have to be for deconstructive ethics to apply to it? According to Jeffrey Jerome Cohen, Heidegger assigns to stone only "agentless perdurance, a blank materiality," but medieval perspectives, and newer philosophies of OOO along with Bruno Latour's actor network theory, allow us to overcome the duality between nature and society. Here, stone "supports, defeats, fosters, yields, impels, risks, resists."[36] This agential nature of lithic and other objects involves relations, and therefore broadly speaking ethical relations. Finally, would an object that entails, or even demands, such relations be called a hyperobject?

This is the nature of hyperobjects in Morton's philosophy. They make a demand on us, and it is an ethical demand, even if the nature of their demand and our inability to respond adequately to them makes us hypocrites. Morton argues that hyperobjects introduce an asymmetry in relation to us, that we end up being confronted by hyperobjects as opposed to confronting them as objects that we can impose our will on and dispose of however we like. According to Morton, "hyperobjects make hypocrites of us all" because anything we do is never commensurate with them and their threat.[37] In the case of global warming, whatever we do only salves our conscience; buying ecological "green" products does not actually change the dynamic of consumer capitalism and the burning of carbon emissions. We do things to make ourselves feel better, like driving a hybrid car, but we are actually hypocrites because at some deep level we know that that is not going to solve the problem. Derrida similarly touches on the unavoidable structural hypocrisy of ethics in *The Gift of Death* when he says that

> As soon as I enter into a relation with the other, with the gaze, look, request, love, command, or call of the other, I know that I can respond only be sacrificing ethics, that is, by sacrificing whatever obliges me to also respond, in the same way, to all the others.[38]

I am doing two things with Morton and Derrida and their relation to Heidegger. First, I am suggesting that Derrida's critique of Heidegger is similar to Morton's rejection of Heidegger's notion of world insofar as Derrida argues that in an ethical relationship, the world disappears, it is gone. And this gone-ness of the world is what allows any ethical relation to occur. Second, I suggest that Morton's focus on hyperobjects and ecology allows us to extend Derrida's deconstruction of the opposition between human and animal to that between living and nonliving beings. By questioning Heidegger's thesis about stones being poor in world, we can think about a theme that Derrida himself does not develop, but one that is developed by an object-oriented-ontology. What Derrida preserves is the explicit ethical responsibility that needs to be a part of our ecological relationships to objects and hyperobjects. And what OOO accentuates is the impossibility of restricting responsible encounters to other human beings or, at most, other conscious animals.

We live at the end of the world, always, not simply after the end of the world. This is because at the end of the world, the world is gone. And this gone-ness of world is the disappearance of a world that allows any genuine ethical encounter to occur. As Derrida argues, death is the end of the world, and mourning constitutes "a world after the end of the world."[39] Death is not just the end of *a* world, but the end of *the* world, which is each time or life unique and irreplaceable. For Derrida,

> Death marks each time, each time in defiance of arithmetic, the absolute end of the one and only world, of that which each opens as a one and only world, the end of the unique world, the end of the totality of what is or can be presented as the origin of the world for any uniquely living being, be it human or not.[40]

Whoever survives is alone and without world, "in a world without world, as if without earth beyond the end of the world."[41] There are endless worlds, and we live at the end of the world each time, all the time, insofar as we are exposed to death and going to die, and at the same time insofar as we have survived others' deaths and have not yet died.

The question is how far we can extend life and death to inorganic objects, and whether the boundary of life marks a limit of genuine relationship. Morton suggests that hyperobjects force us to relate to them differently than we are used to relating to objects, and I am suggesting that an extension of Derrida's thought to Heidegger's first thesis of world provides us a different way of thinking about objects and worlds. If the world is gone in every perishing of an object, then a hyperobject is at once the force field around an object or set of objects that replaces what we used to call an ordered world, and at the same time, what a discrete object becomes at the end of the world.

Religion, understood in Morton's terms, is a massive, complex, and distributed hyperobject. And God, too, is a hyperobject. God does not simply exist in the way that most objects do, or even most hyperobjects that Morton describes. But God inheres or insists at the end of the world because the end of the world allows a different relationship to God as a nonexistent object. This is an alternative relationship of responsibility because hyperobjects, like nuclear waste and global warming, doom us in a way similar to the way that God damns us or judges us in more traditional terms. If objects can be understood as hyperobjects, then I don't see how any object can fail to be a hyperobject in technical terms, despite Morton's preference for large-scale, massive objects in comparison to human beings. Contemporary cosmology considers objects and processes at intergalactic scales and over billions of years, as well as miniscule objects at the subatomic level, including particles that blink into existence for tiny fractions of a second.

I want to supplement this consideration of objects, including hyperobjects, with Lacan's thinking about the object. According to Lacan, whose seminar lecture "On the Names-of-the-Father" I discussed at the end of the previous chapter, human beings are implicated in three realms, or registers, of existence—the real, the symbolic, and the imaginary. The real is first the brute pre-linguistic realm of nonsense that confronts us prior to the advent of language, although later Lacan comes to think about the Real as that which disrupts the symbolic order. The symbolic is shaped primarily by the structure of language, and the structure of language shapes human understanding and desire to such an extent that Lacan argues that our desires are not our own; they are always the desire of the Other. The

Other is a crystallization or concrescence of the symbolic field into a particular site. The Other is an abstraction, but it is necessary for us to understand how symbolic language works. Furthermore, Lacan reads Freud via Saussure's structural linguistics to argue that the unconscious is a broad, intersubjective phenomenon that shapes the symbolic order rather than an individual possession. Finally, the imaginary exists because we always mistake the symbolic for the Real, and this naiveté marks our existence to such an extent that we are always fundamentally mistaking the nature and meaning of our desires.

As Lacan develops his dense idiosyncratic terminology, he suggests that at the level of the imaginary, the big Other that designates the symbolic register is concentrated into a little other, or what he calls in French *objet petit a*, the little other object. The little other is the crystallization of the symbolic big Other that forms a knot around which our imagination gets fixed. The Freudian object *petit a* is the mother's breast, which is the entire existence for the infant baby until she learns to separate the breast from her own body. The little other instantiates the field of human desire at the level of a particular object that warps the entire social field in such a way as to distort it. In his later work, Lacan comes to attach more and more significance to this little other object.

Why this detour into Lacan? Well, I think that we need the idea of the *objet petit a* to really make sense of OOO, and I would like to suggest that it be reformulated as *a*-O-O instead of OOO, or object *a*-oriented ontology. It is important to note that, for Lacan, an object is never a simple object; it is always a strange object, and it confronts us with the limitations of our own symbolic meaning-making as well as implicates us in the Real beyond or within the symbolic. There is a trait that connects us to the object, and this trait goes from the object to us and distorts our perception and understanding in powerful ways.

So in a way, Lacan is already a theorist of OOO before the letter, the first letter, which is *a* for *autre*, or other. Despite his emphasis on language and the symbolic order, Lacan does not propose a linguistic subjectivism any more than Derrida does. His psychoanalytic theory offers profound reflections on the nature of an object, which should inform these newer object-oriented realisms.

In an essay called "Towards a Politics of Singularity," the philosopher Sam Weber reflects on an essay by Walter Benjamin from 1919 called "Destiny and Character." Weber's essay is very complex, but what is interesting is how he draws a connection from Benjamin to Lacan around the notion of the word "trait," or drive, in German *Zug*. According to Weber, this word *Zug* has a dynamic dimension that "tends to be lost in the usual English translation, trait, unless one remembers that every trait has to be traced and describes therefore not a static trajectory but a movement away from something and toward something else."[42] For Weber, this trait is a dynamic character trait, a drive from the destiny of a person to her character, constituting her singularity as a person.[43] The genius of a person is derived from her *Zug*, the drive or trait that connects her to her destiny. This is what makes a person unique, according to Benjamin.

In Lacan's psychoanalytic theory, this *Zug* becomes a singular trait (*trait unaire*) that defines a subject, situating her both in relation to and in some sense outside of the symbolic order. Here, a singular trait is what Lacan calls a signifier, where

> the subject of the signifier, represented by a subject to another signifier, thus inscribed the notion of the *trait unaire* in a network of signification that paradoxically defined singularity as a differential but fully relational notion, a trait or trace, as Derrida would later call it, to stress its temporal relation to what had gone before and what would be coming after.[44]

The signifier indicates the trait or trace of what connects and disconnects a subject to her symbolic language. The key to this notion of the trait is that it proceeds from the object to the subject. It works in the reverse direction from that of stereotypical subjectivism. The object *a* is a strange little object, not simply in its size, but in its ability to establish a trait of character in a subject. This reversal of direction undercuts the correlationism with which Speculative Realism and OOO is so obsessed.

By considering the *a* as a strange object, Lacan helps reverse the relation that appears to go in one unitary direction, from character to destiny or from subject to signification. By closely attending to Benjamin, Weber shows how our character is shaped by our destiny, and the trait or *Zug* derives from this fate. OOO wants to turn around our relationship to objects, but Lacan has already in some ways accomplished this. Here the object *petit a* indicates a dynamic directionality from the *objet petit a* to the subject. The *Zug* or drive is the trait that is traced from the object *a* to the human subject. Every object is potentially an object *a*. And every autre (little a) is every Autre (big A), according to Derrida—he affirms in *The Gift of Death* that *tout autre est tout autre*, or "every other (one) is every (bit) other.[45] Every object *a* is every other object, including a hyperobject in Morton's sense.

According to Morton, we should not reduce objects to the processes that make them up. He says that "a process is just a real object, but one that occupies higher dimension than objects to which we are accustomed."[46]

There is a sense in which a process, a system, and an object or hyperobject can be seen in similar terms, although each has to be qualified to fit into the others. Furthermore, I do not want to pass judgment in ultimate terms and say that an object is a good or bad way to describe reality, although as already explained I do think that these objects are also what Lacan calls *objets petit a*. In any case, if we want to retain this term, we need to see objects as dynamic and changing rather than static and permanent, even if hyperobjects are more stable than we are. For me, energy can be viewed as an object, but that drastically transforms our understanding of objects in the same way that for Morton hyperobjects distort and change what we mean by objects.

For Morton, fate or destiny is named doom, and this doom is delivered to us by hyperobjects. In their ending of the world, hyperobjects deliver doom, where "doom is a decree or an ordinance: a directive." Doom is a kind of judgment as well as discernment, and it can also mean "fate, destiny, and in a stronger sense, death."[47] Our character is determined by the doom of the hyperobject, and according to Morton, this character now consists of hypocrisy. Hypocrisy is "a 'secret doom,'" a pretense or an act. "But it is also simply hidden doom, a message sent from somewhere obscure."[48] As Morton explains, the word "hypocrisy" comes from the Greek term for *delivery*. And this delivery is the singular trait whereby our doom shapes us and renders us hypocrites in relation to these strange objects. Hyperobjects are also objects *a*, and Morton is a little too obsessed with very big objects to notice how great a difference this little *a* makes. He approaches an object *a*-oriented ontology, but does not quite arrive. We can witness the hypocrisy of OOO, but do not simply dismiss it because as Morton affirms, we are all hypocrites. Or as Lacan claims, the non-duped are the ones who are most likely to err ("*les non-dupes errant*").[49]

Hyperobjects confront us at the end of the world, and they force us to reconsider what nature and ecology could possibly mean. Human beings as a species have lived and flourished during a relatively stable climate pattern, but as Michael S. Northcott suggests in *A Political Theology of Climate Change*, "on the current trajectory of greenhouse gas emissions growth, by the end of the present century . . . the planet will be a 'new creation,' but not by the making of God or evolution."[50] The apocalyptic threat of the destruction of human civilization and possibly even extinction of ourselves along with many other animals provides contemporary ecological thinking with a religious or spiritual edge, insofar as we assign spiritual value to living beings. It's unlikely that we could cause the extinction of life on Earth, but human activity can certainly drive, and in fact is driving, many forms of life out of existence. So there is a sense in which we are playing God, wittingly or unwittingly.

I think that to change our fate and avoid our doom we would need to change our nature, which may well be impossible. But we desperately need the impossible. In *Religion, Politics, and the Earth*, Jeffrey W. Robbins and I imagine an impossible synthesis between Hegel, for whom Earth is a substance that becomes subject, and Deleuze, who posits along with Félix Guattari a "geology of morals" in which we are forced to ask "Who does the Earth think it is?"[51] We need to think from the Earth, not just to it, and we need to think Earth itself as a hyperobject without simply assuming that we can remove human self-consciousness from this picture. Morton asserts that Deleuze, like Alfred North Whitehead, is guilty of a "process relationalism" that "conceive[s] time as the liquid in which the image melts and flows."[52] This is an extremely weak reading of Deleuze, who views time in a much more complicated way than as a river in which objects dissolve. I will not discuss Deleuze's understanding of time here, which is treated most explicitly in *Difference and Repetition* and in *Cinema 2: The Time Image*, but I think that there is a real problem with how to think about time in OOO, and there are significant attempts in some of the theorists of OOO to deny and evade the reality of time in order to grant objects a certain timelessness.[53]

Time is entropy, and entropy gives time a directionality to what we understand as time, even though time is not linear. We cannot escape or avoid entropy, although we do need better ways to understand it.[54] The first law of thermodynamics is the conservation of energy. The second law says that all systems or entities tend toward an increase of entropy, measured statistically. This seems to imply an irreversible movement from order to disorder, but then where does the original low entropy state come from? Energy is infinite, it is material, and it is entropic, at universal, physical, biological as well as psychical and metaphysical levels. We are emergent objects (*a*) of entropic-energetic processes, and Earth will survive our deaths, even though Earth will not last forever as a planetary object. We are at the end of the world, and we always have been.

In this chapter, I am reflecting on the current status of Continental philosophy of religion, as well as bringing Derrida's philosophy into closer relation to new materialism and OOO. Much of the time, OOO and speculative realism, as well as some forms of new materialism, represent themselves as opposing what they see as the subjectivism in Derrida and other postmodern philosophies. I am more interested in seeing how some of the poststructuralist theorists like Derrida, Deleuze, and Lacan, who along with Foucault were previously seen as promoting the death of the subject, help us to think about objects and processes in different ways, even as the renewed attention to objects and the discourses of the natural sciences forces us to reconsider the work of Derrida and others.

My guiding thread is to think more deeply about the resonances and the connections of Derrida's famous phrase "*tout autre est tout autre.*" If every other is every other, that does not simply mean all human others who exist in Kantian terms as rational moral beings. Derrida's late work, in particular, has the purpose of questioning the status of the human and its relation to nonhuman others, in a way that has not always been fully acknowledged. Catherine Malabou, whose work I will consider more explicitly in Chapter 7, engages Derrida and deconstruction with a new materialist biological and neurological paradigm. Karen Barad theorizes quantum physics, particularly quantum field theory, with the help of Levinas and Derrida, as I will consider in Chapter 8. But first I will return to Caputo and analyze his understanding of Derrida more fully in the context of Caputo's own philosophy and radical theology.

6

Radical Theology and the Event
Caputo's Derridean Gospel

In this chapter, I focus on Caputo's interpretation of Derrida to show how his attention to the event in religious and then theological terms helps liberate Derrida's philosophy from a specific context of writing. Part of the significance of Derrida's so-called religious turn, which is not really a turn, is an ability to read Derrida's philosophy beyond a straightforward paradigm of writing, and more broadly in religious, ethical, and political ways. Here is the significance of the later Derrida, for better or worse, and we will see how Catherine Malabou understands this in the next chapter. It's not that Derrida himself abandons writing, but that the Anglo-American reading of Derrida is able to loosen up a stricter understanding of Derrida's relation to writing and language. And Caputo's *The Prayers and Tears of Jacques Derrida* is a crucial text for opening up our thinking about Derrida.

It seems a very long time ago now, but Caputo's book, *The Prayers and Tears of Jacques Derrida*, published in 1997, inaugurated a sea change in Anglo-American readings of Derrida. There had already been many readers asking questions about the relationship between Derrida and religion, and as discussed in Chapter 1, many of these questions concerned the connection between deconstruction and negative theology, following Derrida's Jerusalem address, "How to Avoid Speaking: Denials." In terms of philosophy of religion, theoretical religious studies, and postmodern theology, however, these relations invariably took the form of a kind of analogy. Derrida was asking similar questions to those of religion and theology, but he was not himself a religious philosopher or a theologian.

The work of Mark C. Taylor dominated the reception of Derrida in relation to religion in the 1980s and early 1990s. Taylor's *Erring: A Postmodern A/theology* provides an agenda for postmodern theology, with its themes of the death of God, the disappearance of the Self, the end of History, and the closure of the Book. For Taylor, deconstructive writing is a/theological, neither simply theological nor atheological, but in some ways neither and both. Deconstruction subverts theological systems and concepts, although at the same time "the survival of this parasitic discourse presupposes the continuing existence of its host."[1] Theology is the site for Taylor that hosts deconstructive criticism, and deconstruction is a form of "writing that attempts to trace the border and retrace the margin," which "can, therefore, be described as *erring*."[2] For Taylor, then, theology is a form of writing, and Derridean deconstruction is a second-order operation upon writing that traces and retraces the margin of writing, the texts of theology, philosophy, literature, and so on.

As another postmodern theologian, Charles E. Winquist, puts it in his 1986 *Epiphanies of Darkness*, "theology is writing," which means that there is the possibility of a deconstructive theology just as there exists a deconstructive writing.[3] Finally, Carl A. Raschke draws the ultimate conclusion of deconstructive theology when he states that "deconstruction, which must be considered the interior drive of twentieth-century theology rather than an alien agenda, is in the final analysis *the death of God put into writing*, the subsumption of the 'Word' by the 'flesh,' the deluge of immanence."[4] For American postmodern theology, Derridean deconstruction instantiates the death of God in writing. This marginal form of theology is influenced by deconstruction along with the earlier American death of God theology, primarily in the form of Thomas J. J. Altizer's theological writings.

Postmodern theology in the United States elaborates a form of death of God theology along the lines of hermeneutics and then deconstruction as a form of writing, and this is the dominant religio-theological reading of Derrida in the middle 1980s until the middle 1990s. Caputo's book on *The Prayers and Tears of Jacques Derrida* blows this understanding of deconstruction and theology apart, even if it has led to confusion about both Caputo's and Derrida's relation to religion and theology, and this situation eventually forced Caputo to develop his own understanding of radical theology in the first decade of the twenty-first century.

According to Taylor, by the middle of the 1990s, deconstruction had run its course, and Taylor then turned back to Hegel and forward to an interest in networks, technology, and virtual reality. Hegel's philosophy is a better resource, for Taylor, to "think what poststructuralism leaves unthought by showing how nontotalizing structures, which nevertheless act as a whole, are beginning to emerge in the tangled networks and webs through which reality is virtualized and virtuality is realized."[5] Poststructuralism, deconstruction, or the philosophy of difference, on the contrary, "cannot conceive of a structure that does not totalize and is not repressive," while its insistence on difference "has issued in a politics of identity" in which "it is our differences that increasingly are tearing us apart."[6] Taylor's break with Derrida and deconstruction concerns a number of philosophical issues, but it also breaks with a conception of *writing* in Taylor's shift to thinking networks as non-totalizing holistic structures.

Prior to reading *The Prayers and Tears of Jacques Derrida*, Winquist had agreed with Taylor about the exhaustion of deconstruction as a cultural force. However, in his response to Caputo's book on Derrida, Winquist claims that what is crucial to Caputo's understanding of Derrida, and also Derrida's understanding of religion, is the notion of *religion without religion*. Winquist says that "Derrida's claim that he has been read less and less well over twenty years can be understood in the failure to understand his religion without religion."[7] Following Caputo, Winquist distinguishes between Jacques, the philosopher, and Jackie, which is Derrida's actual given name. "Maybe saving the name of Jackie, a task accepted by Jack Caputo, is also a saving of the text that is more than a scholarly exercise sanctioned by professional philosophers," Winquist suggests.[8] What Winquist hears in this work is a new Derrida, an understanding of Derrida's faith as a form of deconstruction beyond writing. Derrida's work is not limited to writing. According to Winquist, what Caputo "has the capacity to hear, when he reads Derrida, [is] that there is always something more or other than the text."[9] And it is the liberation of this "something more or other than the text" in Derrida's work that Caputo's book accomplishes for English-speaking readers who "have been misreading Derrida for twenty years."[10]

This liberation of Derrida's religion without religion as something irreducible to writing marks the significance of Caputo's text. Before turning to *The Prayers and Tears of Jacques Derrida*, I want to briefly consider Caputo's earlier philosophy of religion, primarily his breakthrough work *Radical Hermeneutics*, published in 1987. Caputo's early work was on Thomas Aquinas, Martin Heidegger, and Meister Eckhart, and his 1978 book *The Mystical Element in Heidegger's Thought* demonstrates the affinities of Heidegger's philosophy with Eckhart. *Radical Hermeneutics*, however, is Caputo's breakthrough book, where he begins to really philosophize in his own voice. And it is this growing influence of Derrida on his thought that enables him to do this.

In *Radical Hermeneutics*, Caputo sides with Heidegger and Derrida against Hans-Georg Gadamer, but he keeps the Gadamerian term hermeneutics. Caputo saves the name of hermeneutics in this book, but he radicalizes it. As he explains, here "Derrida is the turning point for radical hermeneutics, the point where hermeneutics is pushed to the brink."[11] Although Caputo turns toward Derrida, he maintains a productive tension between Derrida and Heidegger that he later gives up in *Prayers and Tears*. According to Caputo, "radical hermeneutics situates itself in the space which is opened up by the exchange between Heidegger and Derrida, an exchange which generates a more radical reading of Heidegger and another, more hermeneutic reading of Derrida."[12] The juxtaposition of Derrida with Heidegger produces a "cold hermeneutics," a shiver that also incorporates Kierkegaard and Nietzsche in restoring "the difficulty both to life and to intentionality."[13] This difficulty is the necessary precondition for the production of something new, or a repetition into the future.

In *Radical Hermeneutics*, surprisingly, it is Derrida who is too affirmative, too quick to leave behind this "shudder, this trembling," in which Caputo "locate[s] a cold and comfortless hermeneutics which I think it is the special virtue of Heidegger and Kierkegaard to have expounded."[14] According to Caputo, Derrida, in his emphasis on undecidability, fails to fully open himself to the "mystery" that Eckhart and Heidegger enable us to experience. In the final chapter, Caputo invokes Levinas concerning the face of suffering with which we are encountered, and this experience of the face contributes to "the notion of a more chastened, postmetaphysical notion of religious faith."[15] Attention to suffering alerts us to "the vulnerability of human existence, its lack of defence against the play of the flux," and thus leads us to a religious hermeneutic.[16] Religion is a protest against suffering, although Caputo strips religion of the transcendent power to eliminate suffering. There is a tragic element to human existence, but Caputo does not want this tragic consciousness to be the last word. He wants to dwell upon the importance of laughter as a response to this tragic situation, not as avoidance, but precisely as an affirmation of the difficulty of life. And it is this laughter, humor, and wit that Heidegger misses in his readings of Kierkegaard and Nietzsche.

Caputo affirms a vital significance to the human being. He says that the reason "why I refused to give up on the word hermeneutics" is because of his attempt to "find some way of confronting this question," the question of what it means to be human in a world that does not always comfort or make sense.[17] Later, Caputo does give up the word "hermeneutics," or rather, he replaces the word "hermeneutics" with the word "theology," so that radical hermeneutics becomes radical theology. And with this transition, the mutuality between Heidegger and Derrida that marks *Radical Hermeneutics* tips over into a full-throated affirmation of Derrida. *The Prayers and Tears of Jacques Derrida* is an extraordinary re-reading of Derrida, but it is also a transitional book for Caputo in his becoming-theological, a process that culminates in *The Weakness of God* and *The Insistence of God*.

My claim is that Caputo reads Derrida at first more under the influence of a kind of writing, which is why he needs Heidegger, Kierkegaard, and Nietzsche to supplement Derrida's philosophy with a more existential affirmation of life. But what he sees as he continues to read and think about Derrida is how this existential affirmation is already present in Derrida, even as it becomes more explicit in Derrida's work in the 1990s. Caputo is the keenest reader and interpreter of Derrida in English who is sensitive to this shift and its implications.

In writing about Derrida's prayers and tears, Caputo above all appeals to deconstruction as a *passion*. Passion is not simply writing, a linguistic effect, but more an affect or a force that attests to an otherness of language, writing, and speaking. As Caputo states, "deconstruction is a passion for transgression, a passion for trespassing the horizons of possibility, which Derrida calls, following Blanchot, the *passion* of the *pas*, the *pas* of *passion*."[18] Deconstruction is "a passion and a prayer for the impossible," for what seems impossible or absurd given current standards of logic and norms.[19] We can reasonably expect what is possible, but there is something in deconstruction that desires what is not possible. What deconstruction addresses is *what happens*, which is both strictly speaking possible and impossible in terms of logical conditions of possibility. In his critical philosophy, Immanuel Kant investigates the conditions of possibility for our knowledge of an object. According to Derrida, these conditions of impossibility of knowing something are at the same time conditions of impossibility because our knowing always exceeds these very conditions that make it possible. Deconstruction attends to the impossibility of our knowing and desiring, not just their possibility.

The impossibility at the heart of the possible is a quasi-religious phenomenon. Our desire is for the impossible, for justice, which is not simply or fully possible given the state of the world and its possibilities. This desire for the impossible is never separated from our possible, worldly existence, but in fact it makes existence itself possible, which is a kind of logical paradox or aporia. Derrida coins the neologism *différance* to indicate this dynamic im/possibility at the heart of language and experience that animates deconstruction. Caputo makes it clear, however, that we should not think of *différance* as God, or try to baptize deconstruction as a form of negative theology. Deconstruction, like negative theology, manifests a desire for the wholly other, the *tout autre*, but unlike negative theology, deconstruction understands that the *tout autre* is never found outside of our experience of ordinary others.[20]

Caputo seizes on Derrida's phrase in *The Gift of Death*, where Derrida claims that in certain respects Kant, Hegel, Kierkegaard and Heidegger all "belong to this tradition that consists of proposing a nondogmatic doublet of dogma, a philosophical and metaphysical doublet, in any case a *thinking* that 'repeats' the possibility of religion without religion."[21] Derrida both affirms and distances himself from this specific doublet of religion without religion, but Caputo takes it up and shows how deconstruction is also a religion without religion. For Caputo, deconstruction "repeats nondogmatically the religious structure of experience, the category of the religious."[22]There is a religious structure of experience that deconstruction repeats. This religious structure is not that of this or that determinate religion, but attests to a religion without religion.

The passion of Derrida's thinking and writing, his prayers and tears to which Caputo attends, marks deconstruction as a more-than-linguistic phenomenon, which does not mean that it is simply non-linguistic. Caputo attacks the stupid, reductionist readings of Derrida that proceed by taking literally the translation of Derrida's offhand remark that "there is nothing outside the text" (*il n'y a pas de hors texte*), "as if there *is* nothing other than words and texts."[23] Caputo explains, correctly, that "while there is nothing which, for Derrida, would escape the constraints of textuality, it is no less true that everything that Derrida has written has been directed toward the other of language, toward the alterity by which language is claimed."[24]

In *The Prayers and Tears of Jacques Derrida*, Caputo carefully unpacks many of Derrida's texts from the 1990s, including "Circumfession," *Aporias*, *Specters of Marx*, *The Gift of Death*, *Archive Fever*, and his famous Capri essay on "Faith and Knowledge," to demonstrate where and how deconstruction becomes thoroughly entangled with religion. Caputo provocatively suggests that Derrida presents in these texts his own thinking of a religion without religion, which consists of a messianicity without messianism, a desire for the Messiah or the wholly other that never gets fulfilled. Caputo says that "Derrida too is trying to offer us a work of thought that thinks the structural possibility of the religious, of a certain radical messianic structure, without the dangerous liaisons of the particular religions, without the dogma, without the determinate messianic faiths that divide humanity into warring parties."[25] Caputo isolates, emphasizes, and mobilizes Derrida's religion without religion that comes to the fore in these more recent texts.

Even though this religious reading of Derrida has become much more common, if not commonplace, we should not forget how groundbreaking it was for many English-speaking readers in 1997. Caputo reflects on the historical context of reading Derrida, saying that "in the 1960s *différance* makes a more Nietzschean than Levinasian impression upon us, *différance* looks like the free play of forces, not a way of making ready the coming of the *tout autre*; and one does not detect anything of the prayers and tears of Jacques Derrida or of his religion, about which nobody knew anything."[26] In the 1960s, we did not know how to read or to think Derrida's religion, about which he only later becomes more expressive and more confessional. For Caputo, as for Derrida, religion is a passion for God, but the passion for God is translatable or substitutable with other passions, including a passion for justice, but also a secret for which there is no name.

Caputo concludes that

> Derrida seems to say that "God" is the name of the absolute secret, a placeholder for the secret that there is no secret truth, the blank truth in virtue of which we are always already exposed to multiple interpretations. "God" is a name for the inexpungeable textuality of his life and work, the split in his life that severs him from truth, so that it is up to others to read him (who then know more than him), a limit structure.[27]

God is not a proper name or a master word, but a way to name this unnamable secret. This secret, that there is no ultimate or absolute secret but this lack of an ultimate secret does not thereby eliminate secrecy as such, is the source not only of thinking and writing, but of passion, of prayers and tears.

In Caputo's thought, his reading of Derrida's religion without religion becomes the hinge for Caputo's own development of a radical theology, first explicitly expressed in *The Weakness of God*. In *The Prayers and Tears of Jacques Derrida*, Caputo imagines the possibility of such a Derridean theology when he says: "I have in mind a point at which theology, opening itself to translatability, opens the wound of its own kenosis and suffers from its passion for the impossible."[28] This theology would no longer be able to save God or the name of God from translatability or change, and its desire for God would "fluctuate . . . undecidably with atheism," while its faith "must be faith without faith, without the assurances of faith."[29] This promissory note gets cashed out in *The Weakness of God*.

In the Introduction to *The Weakness of God*, Caputo confesses his weakness for theology. Although he mostly avoids endorsing the term in his previous work, after his move from Villanova University to Syracuse University in 2004, he comes to embrace a weak theology that in *The Insistence of God* becomes a radical theology. Caputo says that he "freely own[s] up here to a certain theological gesture, to a theological desire . . . which is undeniably a desire for God, for something astir in the name of God, a desire for something I know not what, for which I pray night and day."[30] The desire for God is a desire for "something astir in the name of God," which is an event. Caputo proposes that "the name of God is an event, or rather that it *harbors* an event, and that theology is the hermeneutics of that event, its task being to release what is happening in that name."[31] Here theology names the hermeneutics of the event that is sheltered within the name of God, and weak theology attends to this event. Weak theology *releases* the event that is contained in and by the name of God.

For Caputo, the name of God is not a literal name; "the name can never be taken with literal force."[32] The name of God is deconstructible and deconstructed, to allow for the event that it contains to be exposed. Weak theology *deliteralizes* the name of God, while at the same time it does not banish the name of God or prohibit it. Names are translatable, substitutable and deconstructible, but the event is undeconstructible. Why? Because it is not a thing; there is nothing substantial to deconstruct. The event is a happening, but it is not simply what happens in an obvious or literal way. As Caputo paraphrases Gilles Deleuze, the event is what is going on in what is happening. The event is a singularity, it is a difference that makes a difference, and it is a fundamental transformation or metamorphosis of a situation.

Caputo seizes on the significance of the concept of event for poststructuralism, mainly Derrida and Deleuze, and he develops first a hermeneutics and later a theology that would be responsive to and expressive of the event. The event is not just an event of writing, a linguistic event; it is no less a religious and theological event. Caputo says that "the event that is promised by a given name is what Derrida calls 'the undeconstructible.' The event is always undeconstructible because it is always promised or called for, always to come, whereas what actually arrives has arrived under present conditions and is always deconstructible."[33] What Caputo recognizes is that the name of justice in Derrida's essay "Force of Law" is an event, and that is what makes it undeconstructible. The event is the messianic horizon of every action and being, but the event is also the surprising and unexpected aspect of experience that breaks with every horizon and exposes any action or being to what it is not and what it never expected.

Caputo contrasts his weak theology with a strong theology that holds fast to a literal name and understanding of God. "In a strong theology," he writes, "God is the overarching governor of the universe, but in what follows I will endeavor to show that the weak force of God settles down below in the hidden interstices of being."[34] We have to be careful not to simply oppose weak to strong, as if they were opposites. Weakness is not lack of strength; it is nothing at all relative to strong theology. Or rather, it is only something insofar as it is a call or a promise, a provocation or a charge. From the standpoint of determinate religion, God or god is a strong force, a power to do something substantial in the world. Caputo is not invested in this traditional form of God. He is interested in thinking about God from the perspective of religion without religion, which means that God is not a thing, a person, or a power. God is that name that names for certain religious people something transformative, and it expresses an event, even while covering up the phenomenon of the event to a certain extent. Caputo is not interested in God as such; he is interested in the event that the name of God shelters. Just as Caputo liberates Derrida's deconstruction from its textuality by demonstrating its religious passion, in *The Weakness of God*, Caputo applies deconstruction to theology in order to liberate the event from its secure place within what we sometimes call God.

In *The Weakness of God*, Caputo proposes a God without sovereignty, in line with my perspective in Chapter 3 that Derrida's philosophy implies a political theology without sovereignty. God is without sovereignty, and God is without being in any traditional sense. When we read Caputo's theology, the danger is to read it in conventional theological terms, whether we are atheists or believers, and think that God must be some sort of hyper-being. This is the whole point of Caputo's appropriation of Derrida, however. It is absolutely not the case that deconstruction purifies God for us so that we can affirm a God beyond being in the way that Jean-Luc Marion does.[35] No, God is not something, somewhere transcendent. God is not anything, but the word God names a call, a promise, that is the event of justice. The call of God as the event astir in the name of God keeps the world from closing in on itself, not because it is opened up by some transcendent other, but because it is an immanent dynamic of constitution and in-constitution that Derrida names *différance*, and then *khōra*.

Caputo spells out what he does not mean by "God" very clearly in *The Weakness of God*:

By "God," on the other hand, I do not mean a being who is there, an entity trapped in being, even as a super-being *up there*, up above the world, who physically powers and causes it, who made it and occasionally intervenes upon its day-to-day activities to tweak things for the better . . . That I consider an essentially magical view of the world . . . I mean a call that solicits and disturbs what is there, an event that adds a level of signification and meaning, of provocation and solicitation to what is *there*, that makes it impossible for the world, for what is there, to settle solidly in place, to consolidate, to close in on itself.[36]

The name of God contains a promise, a call to fashion the world otherwise than it is, and an invitation to participate in this ongoing creation of signification and meaning.

The Weakness of God explores biblical narratives and themes with an eye to transforming them according to another "logic of sense" than the one that they have been given by strong theologies. Caputo asks the question about these influential texts: "What event do these stories harbor? What do these stories mean? Hermeneutics is all. All things flow in a river of meaning."[37] In his book, he defamiliarizes these stories by reading them through Deleuze's reflections on the paradoxes of Lewis Carroll in his book *The Logic of Sense*.[38] This alternative logic of sense takes place in what Jesus calls the kingdom of God, but the kingdom is not a literal spatial reference. The kingdom, as expressed by and through Jesus, "is the locus of divine transformation" where "things are remade, refashioned in accord with their origin and congenital goodness."[39] Caputo distinguishes between miracle and magic, where magic refers to the literal coming back to life of Lazarus or Jesus, while a miracle is a theological term that "harbors an event of a deeply incarnate kind."[40] Salvation and resurrection are not magical fantastical occurrences, but genuinely transformative experiences that testify to the event that Jesus released in his preaching of the kingdom of God.

Caputo argues that "salvation is situated, not in a heavenly pleasure but in the pain of the present, 'the very instant of pain.'"[41] Citing Levinas, Caputo says that messianic hope is not simply for the future, but for the future in the present, for the new beginning of what is taking place right now. The solitary ego imagines its own afterlife as an immortal existence into the indefinite time of the future. But that's not what salvation and resurrection mean, at least not in terms of an event. We want to preserve our own being from the full consequences of the event, the fact that we are not who we are, and cannot fully close in on ourselves in a gesture of auto-immunity: "Neither time nor salvation, neither rebirth nor resurrection, is possible in the solitary ego."[42] For Caputo and for Levinas, messianic time refers to the coming of the other, which means the hope for more time and for forgiveness because we are not just ourselves but are already other. The relief of suffering is not simply the end of suffering. It is "just this suffering for which nothing can compensate that constitutes the 'torsion' and 'exigency' of the moment" and what "gives it the force or energy to 'unleash the future,' to open up the future and make a new beginning possible."[43]

The Weakness of God elaborates a theology of the event. It does not refuse the name of God, but it refuses to be tethered to what this name ordinarily means. The weakness of the title is not a weakness that could be contrasted with strength on any linear continuum, but a twisting free from strong theology and strong religion. Weakening theology involves deconstructing theology, but deconstruction is not a negative procedure. Deconstruction is an affirmation, as Derrida claims and Caputo well knows, and it releases the event from its name without thereby abolishing the name.

Another way to speak of God is to say that God does not exist, but rather God *insists*. In *The Insistence of God*, Caputo develops a theology of "perhaps," which is a term that Derrida often uses. The insistence on the word "perhaps" underlines the undecidability that is irreducible to religion, faith, and God for Caputo. He says that "something is calling, or rather something is getting itself called, in and under the name of God, of 'God—perhaps,' inasmuch as the caller in the call is structurally inaccessible, unidentifiable."[44] The event that Caputo associates with the name of God is a *call* to justice, ethical action, and responsibility. God is not a thing or a being who could exist; God is the inscrutable name that indicates the source of this call to responsible action. God's call insists upon our lives.

Caputo explains that his faith "is placed in what is going on *in* the name (of) 'God' and of 'theology,' which is the insistence of the event, or the chance of the event, and the corresponding faith that God can happen anywhere."[45] His faith is not simply faith in any determinate person or concept of God, but "a deep and structural faith" in the possibility of an event that can transform us and make us better. The "perhaps," however, is the fundamental acknowledgement that an event might not be good or make us good. It might be awful and terrible; it might constitute a disaster. Or we might ignore the call and cling to our comfortable habits and living and thinking. Following Derrida, Caputo affirms a radical hospitality, where hospitality "is a figure of the event" that signifies a welcoming of the other, while acknowledging that the in-coming of the other might not be a good thing.

Caputo spends less time in this book on biblical stories than he does in *The Weakness of God*, but he does draw out a contrast between Mary and

Martha as two orientations to the world and the possibility of an event. In Luke 10:38–42, Jesus visits the home of Martha and Mary, and praises Mary for devoting her attention to his teachings while implicitly chastising Martha for complaining that she is not getting any assistance with her domestic work. Caputo follows Meister Eckhart's reversal of the traditional reading of this passage that favors Mary's spiritual contemplation over Martha's worldly action. For Caputo, "Martha is an emblem for me, a figure in whom all the dynamics of the event, of the insistence of the event, are contracted." Martha is emblematic because Martha does not simply attend to the spiritual needs of Jesus and others but she is primarily focused on meeting material human and animal needs in her performance of hospitality.[46]

According to Caputo, radical theology affirms the insistence of God over the existence of God, pays attention to the radical demand of hospitality, and stays with the ambiguity of the "perhaps" all the way. Radical theology in this book names what Caputo calls "weak theology" in *The Weakness of God* and "radical hermeneutics" in *Radical Hermeneutics*. Radical theology is derived from more orthodox, traditional, and confessional theologies, but it distorts and deforms them by being more faithful to the insistence of the event and the irreducibility of the perhaps. "The confessional theologies are the only theologies that exist," he writes, "while radical theology, which does not exist, insists or haunts the confessional theologies."[47] Radical theology breaks with the boundaries and authorities that circumscribe confessional theologies. Radical theology "reserves the right to ask any question, without regard to whether it fractures or divides the community or causes schismatic conflict and confessional breaks or engages in revisionist readings of classical scriptures."[48] Radical theology is radical all the way down to the roots, and it answers not to this or that particular institution or authority, but affirms a "hermeneutic universality" that strives "to talk to anyone, anywhere, anytime."[49]

Caputo contrasts a species of postmodern theology or philosophy of religion derived from Kant that tries to limit the bounds of reason to make room for faith, with a more radical form of postmodern theology/philosophy of religion that is influenced by Hegel. The post-Kantian form of philosophy of religion is more epistemological and apologetic, striving to defend a realm of faith freed from the attacks of modernist rationalism, secularism, and atheism. This is a valid endeavor, but it ultimately tames postmodern philosophical and theological thinking because it contains it. The Kantian version of postmodernism is more of "an abridged postmodernism" that tempers the absolutism of religious believers and atheist nonbelievers alike.[50] "In the version that descends from Hegel," however, "postmodern theology is neither an epistemology nor an apologetics but a genuinely radical theology which mounts a heartier critique of confessional two-worlds theory."[51] There remains a residue of implicit dualism in Kantian and post-Kantian philosophies of religion due to the distinction Kant makes between the noumenal (the thing in itself, which is transcendent) and the phenomenal (the thing as it appears to us, which is immanent) in his work.

Hegel overcomes what he perceives as a Kantian dualism, and Caputo affirms this aspect of Hegelian thought, despite his reservations about where Hegel ends up, which is an affirmation of absolute Spirit conceptualized in terms of the Concept or Notion, *Begriff*. Caputo explains that he affirms a heretical Hegelianism that disavows the teleology of philosophical *Begriff* and the progression of absolute Spirit that dominates Hegel's philosophy. Caputo says that "what I am calling a theology of the insistence of the event is a heretical version of Hegel, a variant postmodern Hegelianism, a kind of hybrid or even headless Hegelianism without the Concept."[52] Caputo asserts that Hegel rather than Kant is the true father of radical theology, and he affirms Hegelianism in its heterodox form.

Later in the book Caputo distinguishes his Hegelianism, still strongly influenced by Derrida, from the Hegelianism of Catherine Malabou and Slavoj Žižek. He also offers some insightful engagements with the newer philosophy of Speculative Realism. One question about Hegel is just how orthodox Žižek's and Malabou's interpretations are, and if they are also heretical, how much they are similar to or divergent from that of Caputo. I think that they are closer than many readers would suspect. By way of a conclusion to this chapter, as well as a transition to the next chapter that will focus on Malabou's relationship to Derrida, I want to focus in more closely on Caputo's critique of Malabou's understanding of Hegel.

In Chapter 6 of *The Insistence of God*, a short but incredibly profound engagement with the philosophy of Catherine Malabou and her reading of Hegel, Caputo drops a bomb. The chapter title, "Is There an Event in Hegel?," attests to the significance of Caputo's reevaluation of Hegel in the middle of this book, which as we have seen is an affirmation and appropriation of Hegel for a radical theology of "perhaps." This affirmation, however, can only go so far, and thus Caputo is forced to clarify and delimit his interpretation of Hegel in relation to those of Malabou and Žižek (he criticizes Žižek in a later chapter).

So the answer to the question of whether there is an event in Hegel is a kind of "perhaps," which here for Caputo does not mean undecidable; it means yes, up to a point, but ultimately no, not a radical enough event of the sort elaborated by Derrida and affirmed by Caputo. And this chapter gets at the heart of what's at stake between radical theology and contemporary Continental philosophy in relation to the readings of Hegel,

Heidegger, and Derrida. Caputo advocates a heretical Hegelianism, and he claims that Malabou's Hegel is also heretical but she is perhaps not explicit enough about this heresy, insofar as she reads Hegel through Heidegger but fails to admit it. Caputo says that Malabou's speculative hermeneutics concerns the plasticity of "the auto-transforming life of the Absolute in time."[53] The Absolute is another name for Spirit, which is the subject of Hegel's philosophical narrative. And the stakes are to what extent Malabou can read a radical contingency or accident into the necessity of essential Spirit.

On the one hand, Caputo argues that there is contingency in Hegel's dialectic of Spirit; it does not know how it will unfold in time. We cannot see what is to come. On the other hand, Caputo claims that at the end of the day, at dusk, we can see what has come, what has happened, and can declare a retroactive necessity. Caputo's central "claim is this: nothing is going to happen that does not fulfill the destination of Spirit. If 'eventually' the Spirit can see these unforeseeables coming, this undoes the 'event.'"[54] Because Malabou's argument depends on Hegel's, Caputo can only go so far with Malabou. Malabou's plasticity is tied to Hegel's, and this limits the chance of the unforeseen event. An event can surprise us within a certain range or framework, but it cannot explode the framework itself. Caputo concludes that "there is no absolute errancy in Hegel, no absolute waste, no errancy that reaches as far as the absolute itself."[55]

I have to confess that I do not know whether or not this is a correct reading of Hegel, to the extent that it would supersede or render incorrect these important contemporary readings of Hegel by Malabou and Žižek, as well as that of Katrin Pahl in her important book *Tropes of Transport: Hegel and Emotion*.[56] I am not an expert or confident enough reader of Hegel to declare or decide that Caputo is right and Malabou is wrong, or vice versa. It's possible that Malabou's reading of Hegel is impossible (and it is certainly only possible via Heidegger), but I don't think it simply conforms to the strictures of Caputo's understanding and presentation of Hegel in his chapter.

Caputo lays out and endorses Derrida's reservations about Malabou's Hegel, and they are the same reservations that Derrida articulates in his Preface to Malabou's book *The Future of Hegel*. Derrida's Preface is called "A Time for Farewells."[57] Caputo focuses on Derrida's question about the death of God, and how radical it is in Hegel's philosophy. As Caputo says, "could God, unawares, step on an explosive? Could God be blown to bits without so much as knowing what hit him?"[58] Such an event would be an accident, and Derrida suggests that Hegel's philosophy cannot make room for an absolute accident that would explode Spirit itself. In his Preface,

Derrida says that Hegel could never subscribe to a history in which God or Spirit could accidentally blow up:

> A God who would have, without ever seeing it come, let an infinite bomb explode in his hands, a God dead by some hopeless accident, hopeless of any salvation or redemption, without essentializing sublation, without any work of mourning and without any possible return or refund, would that be the condition of a future, if there must be such a thing called the future?[59]

In Hegel's dialectic, according to Derrida and Caputo, there is no chance that God could be blown apart by an infinite bomb, and this is a delimitation of the event as well as a limit of plasticity in Hegel.

I want to make two points here in response to Caputo's channeling of Derrida's questioning of Malabou's Hegel. First, my reading of Malabou and Malabou's philosophy as she develops it after *The Future of Hegel*, partly in response to Derrida's critique, suggests that plasticity is an event insofar as explosive plasticity is articulated along with the other two characteristics of plasticity, the ability to give form and the capacity to receive form. There is an event of plasticity in Malabou, and this becomes clear in her work on brain plasticity as well as her powerful readings of Freud in *The New Wounded* and Heidegger in *The Heidegger Change*. Just as Caputo suggests that there cannot be a thinking of accident without the notion of essence, so there cannot be the idea of force without form, and furthermore, we do not know what it means to have or think an event except in contrast to some sort of being or structure.

I think that Malabou suggests that for Hegel, Spirit *is* this errancy and waste, that it is not a circular process of Spirit becoming itself but an originary metamorphic change that we call Spirit afterward, in hindsight. It's not that Spirit cannot die or that there is any limit to what can happen to Spirit by accident; it's that whatever happens can only be affirmed or imagined to be Spirit essentially so long as there is subjectivity to think it. So the question is, can Spirit die? Of course it can, it does all the time, and this explosivity of and to Spirit constitutes Spirit; it "makes" Spirit in us, it makes us inspire and expire. Spirit is change, exchange, metamorphosis. Death is not something that occurs in the future, just as for Derrida the future is not simply the indefinite extension of the present. According to Caputo, in Hegel "the essential form does not mutate," but in Malabou mutation is the "essence" of form.[60] For Malabou, Derrida's messianicity of the event threatens to swallow up form and induce a passivity into philosophy that she turns to Hegel and to brain plasticity to undo. Plasticity would be this forming of a future that we cannot fully form, but we can take responsibility for participating in and shaping it. I don't think there is a teleology inherent in plasticity, although I do struggle with the apparent teleology in Hegel at the level of his writings.

My second point is to emphasize just how explosive Caputo's theology is here in this chapter and generally. According to Caputo,

> for there to be a future for God, God would have to be exposed to the final and uttermost risk of death, where death would be something more than a moment in a metaphysical transition, more than the plasticity of transformability, but the possibility of extinction, of entropic dissipation, of a thermal equilibrium overtaking the divine fire, where there would be neither form nor transformation, where the logic of the dialectic would be exploded by the logic of death and utter irreversible extinction.[61]

If there is an absolute and irreversible extinction as speculated by Ray Brassier in his provocative book *Nihil Unbound*, there is not only no more God, but no more form, and therefore no extinction of form. Caputo does not deny Brassier's challenge; he acknowledges the "logic of death and irreversible extinction." This understanding operates dialectically in Caputo's theology, to give rise to further forms of thought and practices of life for Mary, for Martha and for us.

Caputo says that Hegel and the theologians are on the same side in opposing this logic of death and irreversible extinction, and I want to underscore the radicality of this thought, this radical theological thought of the death of God at the heart of *The Insistence of God*. This thought is explosive, and it breaks with most recognizable forms of theology. The question is whether this explosive theological thinking is, in fact, plastic, in Malabou's sense, and also whether in some sense Malabou and Caputo are on the same "side." Malabou is not a theologian. She does not want to hold onto the life of God, or save God from risk of death. Perhaps plasticity is incompatible with a weak theology of the event, at least from the viewpoint of conventional philosophy and confessional theology. But for "a new species of theologians," it might not be possible or necessary—or even in the last instance accidental—to choose between plasticity and the event, between Caputo and Malabou, or between two futures of Derrida.

7

Deconstructive Plasticity
Malabou's Biological Materialism

As we saw in the previous chapter in Caputo's critique of Malabou, Derrida expresses his reservations about Malabou's understanding of the plasticity of Hegelian Spirit in his Preface to her first book, *The Future of Hegel*. Derrida's Preface is called "A Time for Farewells," and he asks whether one could ever finally bid farewell to Spirit, or would it necessarily always return?[1] In this chapter, I explore Malabou's development of the destructive plasticity of being, form, or Spirit, partly in response to Derrida's critique. Malabou's understanding of plasticity accomplishes two things: First, she argues that plasticity replaces writing as a motor scheme, and second, she advocates a biological materialism that remains faithful in some respects to the legacy of deconstruction.

Although Caputo and Malabou are faithful to Derrida in very different ways, I am suggesting that they are both reading Derrida from beyond the vantage point of writing as a motor scheme. Malabou does not share Caputo's Derridean faith, and she is much more critical of Derrida's later work, but she does so out of an effort to graft plasticity onto writing as the shift from an earlier to a later Derrida. Caputo, as we have seen, attends to this religious or messianic passion to which Derrida increasingly gives voice, and he situates religion without religion in the context of a broader economy than other English speaking readers. Caputo over-reads Derrida's religious passion, perhaps, but he is right to see, to celebrate, and to liberate it. Malabou does not share this religious passion, but her desire for plasticity can be seen as complementary to Caputo's reading of Derrida's faith.

As already mentioned at the end of the previous chapter, I do not think that Malabou's notion of plasticity simply fits into Caputo's critique, which follows Derrida's response to her interpretation of Hegel. At the same time, I think that Malabou develops the negative or destructive aspect of plasticity partly in response to Derrida's critical questions in his preface to her Hegel book about whether Spirit could explode. Here again I want to distance my interpretations of Caputo, Malabou, and Derrida from any exegetical questions about Hegel's texts. First, I will focus on Malabou's contrast between plasticity and writing as a motor scheme, and then I turn to her elaboration of the destructive elements of plasticity. Finally, I will show how Malabou's interest in biological plasticity generates a new materialism.

In her book *Plasticity at the Dusk of Writing*, Malabou reflects on the philosophical trajectory of her entanglement with Derrida, Heidegger, and Hegel to her breakthrough work in theorizing the brain sciences. She says that Derrida's concept of "arche-writing" represents an enlargement and a modification of writing, and that this enlarged conception of writing functions for Derrida and other philosophers as a motor scheme. This scheme is connected in an important way to temporality and history. Malabou says that "all thought needs a scheme, that is, a motive, produced by a rational imagination, enabling it to force open the door to an epoch and open up exegetical perspectives suited to it."[2] Writing for Derrida becomes a generalized motor scheme, sufficient to explain and justify deconstructive readings of texts.

Arche-writing is the object of a new science of grammatology that liberates writing from language and linguistics in any narrow sense. Writing attends to "the *general movement of the trace*, the original breach without which speech would be impossible."[3] So writing is never simply what we literally think of as writing because it becomes much more generalized. Malabou argues that plasticity can be seen at first as yet another modification of writing, including by Derrida himself in his Preface to *The Future of Hegel*, but she comes to view plasticity as an alternative and successive motor scheme. In her book on Heidegger, *The Heidegger Change*, Malabou understands Heideggerian ontological difference as change or metamorphosis. Change is fundamental, and the presupposition for thinking anything at all. Plasticity then becomes a change or modification of writing, but one that is radical rather than derivative. Plasticity is not simply a modification of writing, but a transformation of writing into plasticity as a new motor scheme.

Malabou states that "a motor scheme, the pure image of a thought—plasticity, time, writing—is a type of tool capable of garnering the greatest quantity of energy and information in the text of an epoch." This pure image of thought, a Deleuzian term originally from *Difference and Repetition*, names what Deleuze calls a plane of immanence in his and Guattari's work. Writing can be seen and named as a pure image of thought in its time now that that time is passing; it is at dusk, "the dusk of *written form*."[4]

If plasticity is now the motor scheme that replaces writing, it is because "the concept of plasticity is becoming both the dominant formal motif of interpretation and the most productive exegetical and heuristic tool of our time."[5] Plasticity is more adequate to the biological and neurological sciences of our time than writing is. Writing became a motor scheme during an epoch "that began with structuralism and found its mooring in linguistics, genetics, and cybernetics."[6] In the twenty-first century, the prevalence of the brain sciences changes how we think about thinking because we no longer operate with a simple opposition between form on the one hand and gap or trace on the other. All of these concepts transform themselves in relation to what we are able to learn and discover about cerebral plasticity, such that we no longer work with graphic metaphors but rather "*assemblies, forms, or neuronal populations*."[7] Writing cannot explain how the brain works, but plasticity can and does.

Writing contains a graphic element that is irreducible, and the notion of writing as a motor scheme necessarily implies an absent/non-absent trace. The trace defines writing as a motor-scheme because writing consists of leaving and then interpreting traces. Traces are always traces of something. The incompleteness of traces is what is changed in the transformation of writing into plasticity. Plastic traces are not signs of something else, but forms-in-formation, including transformation and annihilation of form itself. In another essay, "Grammatology and Plasticity," included in her book *Changing Difference*, Malabou elaborates on this delimitation of Derrida. She explains how grammatology is not a strict science, but it takes on a certain scientific form, the impossibility of a "science of writing," to illustrate what she calls in *Plasticity at the Dusk of Writing* a motor scheme.

Derrida names arche-writing as "the original trace, the deferment of presence and of living speech" to suggest a "generalized writing that 'covers the entire field of linguistic signs,' in other words, the entire field of human activity."[8] Malabou argues that this modification of writing is still a re-writing, and takes place under the sign of writing in general. But what happens if and when writing comes to an end? Plasticity is what happens "after" writing for Malabou. She asserts that "it is clear today that writing, as motor scheme, is no longer pregnant in the real."[9] In *Plasticity at the Dusk of Writing*, Malabou claims that "*plasticity is the systemic law of the deconstructed real*, a mode of organization of the real that comes after metaphysics and that is appearing today in all the different domains of human activity."[10] The real as a locus of the motor scheme has a Lacanian resonance, and we can also recall that Alain Badiou characterizes the twentieth century as having a passion for the real.[11]

Deconstruction always relates to the real, whatever its motor scheme, and Malabou philosophizes beyond Derrida and beyond deconstruction because she questions writing as a motor scheme. Malabou replaces writing with plasticity as a substitute motor scheme, and I am not arguing that she is simply right or wrong, although her work is extremely convincing that plasticity is a new motor scheme, even if it is not entirely clear whether or not it is *the* motor scheme, or whether there can be only one at any given time. I am suggesting that something changes, and Malabou is attuned to this transformation, and she attempts to think Derrida's philosophy in this new mode. Furthermore, even if Derrida himself refuses plasticity as a motor scheme, my argument is that this transformation is what drives Derrida to think and write deconstruction differently after 1989.

After 1989, Derrida comes to express his ideas more explicitly in terms of a paradoxical relationship between technics or the machinic, and a kind of ethical responsibility as openness to the other, including the other in me. The context for this twofold reflection is less writing in any explicit or even general sense, and more ungrounded. Malabou offers a ground, or at least a scheme, for us to help think through and beyond Derrida's own philosophy. In constituting her understanding of a motor scheme by means of neurological plasticity, she wants to close the gap between technics and responsibility that Derrida wants to hold open. Derrida works with and through the paradoxical tension between the machinic repetition and the singular dignity of life as ethical responsibility to the other, whereas Malabou wants to unify both in her conception of plasticity. Derrida's later work thus appears more ungrounded, and he would resist adopting Malabou's characterization of this new motor scheme, but her idea of plasticity gives us a vantage point from which to make Derrida's philosophy more coherent, even if it betrays some of the letter of his writings.

Malabou does not follow the religious implications of Derrida, as Caputo does, but both are attentive to something that perhaps Derrida was not fully aware of, that at a certain moment in time deconstruction twists free of writing. This liberation has religious implications, but it also has political, scientific, and other implications, whether we want to follow or endorse them or not. The question that drives this book is whether Derrida's philosophy has a future, and its tentative suggestion is that this answer depends on the extent to which it can be released from writing. According to Malabou, "the choice seems simple: either we recognize that deconstruction is dead and repeat that this is the case, or we accept the new change in modification, in other words, a change of difference."[12] This change of difference names the exchange of plasticity for writing, and it names a future for deconstruction and for Derrida, at least for Malabou.

Much of the possibility of envisioning plasticity in the wake of writing lies in plasticity's destructive power. According to Malabou, plasticity involves both the capacity to receive form and the capacity to give form, from the Greek word *plassein*. These two complementary aspects of plasticity in a classical sense are supplemented by the third capacity of plasticity, "the capacity to annihilate the very form it is able to receive or create."[13] This explosive quality of plasticity involves the auto-annihilation of form. This explosive annihilation of form is necessary for repair, for healing, and for growth of neuroplastic cells, according to Malabou. In *What Should We Do with Our Brain?*, she argues that "the sculpture of the self is born from the deflagration of an original biological matrix, which does not mean that this matrix is disowned or forgotten but that it cancels itself."[14] Plasticity takes place between shaping of form and destruction of that form itself. Destruction of form is an intrinsic part of the process of formation.

In *The New Wounded*, Malabou develops a critique of Freud by suggesting that the sexuality of the unconscious mind is changeable but not destructible. The limits of a psychoanalytic understanding of the person lie not in emotional mental trauma, which can always be recuperated into the existing self, but in brain injury. Brain injury changes the person so profoundly that we cannot simply say that it is the same person. Malabou's careful reading and delimitation of psychoanalysis should not be seen as a crude biological reductionism, but a way to challenge the presuppositions and limits of Freudian and Lacanian psychoanalysis itself.

The New Wounded is important for many reasons, but primarily for the distinction Malabou makes between the psychoanalytic notion of sexuality and the neurological idea of *cerebralité*; that is, the understanding of how the brain works changes how we conceive of an event. The cerebral event radically transforms subjectivity, while the sexual event is always assimilated into or appropriated by the subject. What Malabou is interested in here is the destructive plasticity represented by brain wounds, whether caused by trauma as in post-traumatic stress disorder or diseases such as Alzheimer's. This destructive ability of brain wounds to profoundly and irrevocably alter the self makes it entirely different from Freudian psychoanalysis, which always incorporates external events into internal, psychic and sexual processes, whether conscious or unconscious. She claims that "the resistance of cerebrality to sexuality, in the final instance, pertains to the manner in which the cerebral self *belongs to the other without alienation or specularity.*"[15] Neurological discoveries expose the contingency and fragility of identity, which Malabou then draws upon to show how these processes change how we have to think about Freud.

Malabou draws on current neurological research and contemporary psychoanalytic works and applies them to a careful, penetrating and convincing reading of Freud's primary texts, in order to fashion her original interpretation. She claims that Freud ultimately fails to get beyond the pleasure principle, despite his later intentions, because he always reduces events to internal sexual causes, and he cannot truly think the possibility of external chance or accidental events. The psychic or sexual event is the appropriation of any event whatsoever into the psyche, and this linkage forms a totality in Freud's thought. On the other hand, Freud cannot think a purely cerebral event, one that comes from outside and cannot be mentally connected or assimilated into a subject's psychic processes. What is so interesting and ironic, of course, is the fact that the brain is seen as "internal" in bodily terms, but its wounding or alteration is inassimilable into psychic relationships. Brain wounds so radically alter personality that someone can become someone else, and this is a loss so total that it precludes mourning, except by others.

At the end of *The New Wounded*, Malabou rewrites the Freudian death drive in cerebral or neurological terms. The death drive is beyond love and hate, sadism and masochism, because it is associated with the cerebral event, the destructive annihilation of personality by means of a wounding trauma. The death drive is the augur of a new materialism, a materialism that is completely outside the psychic subject, and the subject is exposed to a vulnerability that she cannot control or assimilate. She claims that "only profound reflection upon destruction, death, and the negativity of the wound will make possible a truly efficacious and pertinent approach to the neuropsychoanalytic clinic."[16] Although her reading is a critical reading, Malabou does not simply dismiss Freud's work and significance, or claim that neurological research makes it obsolete in a straightforward scientific or positivistic way. By re-writing the death drive from the standpoint of the cerebral event, she forces readers to confront and engage with Freud and post-Freudian, including Lacanian, thought in a different and important manner.

In *Self and Emotional Life*, coauthored with Adrian Johnston, Malabou further reflects on the theme of destructive plasticity that she has elaborated in *The New Wounded*. In her contribution to this collaboration, she focuses on the question of affect, and engages what Derrida calls "heteroaffection." Affect for Derrida is always heteroaffection, and deconstruction shares with psychoanalysis a focus on affect as predominantly related to the subject, and loss of affect as alienation. In this model, "the loss of affects is . . . the subject's total disconnection from her affects."[17] In her contribution in *Self and Emotional Life,* "Go Wonder," Malabou focuses mainly on Descartes and Spinoza, and she also considers the writings of the neurologist Antonio Damasio. She ties this reading, as she does so much of her work, to a critique of deconstruction. Malabou explains that "one of the major points of discussion between philosophy, psychoanalysis, and neurobiology concerns not only the possibility of heteroaffection, but the possibility of a hetero-heteroaffection."[18] Heteroaffection is still an affect, but hetero-heteroaffection breaks with any sense of affect and destroys the foundation of our sense of what it means to be a self.

Hetero-heteroaffection in "Go Wonder" plays a role similar to cerebrality and the possibility of brain injury in *The New Wounded.* Her argument with Derrida in "Go Wonder" parallels her critique of psychoanalysis in *The New Wounded.* She turns to Damasio for a thinking of the self that does not presuppose a baseline psyche. "The neurobiological approach to emotions," Malabou suggests, "allows us to think *a strangeness or estrangement of the self to its own affects.*"[19] The self can only be thought, as it can only be fashioned, in negative terms, according to this destructive plasticity, which "forms and sculpts a new identity."[20] According to Malabou, destructive plasticity does not simply destroy. It's also forms something new, even if this something or someone is so radically different as to make recognition impossible. The result of destructive plasticity in the form of a serious brain injury "is the formation of 'someone else,' a new self, a self that is not able to recognize itself."[21]

Destructive plasticity marks an extreme limit of plasticity in its negative form, but it still manages to contribute to a new formation. Destructive plasticity incorporates what Derrida analyzes as the machinic repetition of technics in "Faith and Knowledge" and elsewhere. The productive promise of plasticity also generates a kind of ethical responsibility, even as it denies the transcendence that is usually associated with Derrida's and Levinas's ethics. How does this work? In an essay from *Changing Difference,* "The Phoenix, the Spider, and the Salamander," Malabou responds more explicitly to Derrida's critique in "A Time for Farewells" by offering an interpretation of Hegel's sentence from the *Phenomenology of Spirit*: "The wounds of the Spirit heal and leave no scars behind."[22] We can read and interpret this sentence in at least three ways: dialectically in a conventional sense, deconstructively in a Derridean sense, or post-deconstructively, which is the reading that Malabou wants to suggest. She fastens on the example of the salamander, which heals its amputated tail without leaving a scar due to specialized trans-differentiated stem cells. "When a salamander or lizard's tail grows back," she explains, "we do indeed have an instance of healing without a scar. The member reconstitutes itself without the amputation leaving any trace."[23] This example is crucial for Malabou because here destructive plasticity works to regenerate without leaving a scar or a trace. Plasticity works in a way that deconstruction, according to the scheme of writing, does not.

Plasticity is destructive, but this destructive nature of plasticity is not simply negative. It is also metamorphic. The third element of plasticity, its explosive aspect, changes radically, so radically that our presumptions of identity may no longer hold. But this radical plasticity is also a form of regeneration and freedom because it is "only in making explosives does life give shape to its own freedom, that is, turn away from pure genetic determinism."[24] An energetic explosion is the idea of nature, a Hegelian Idea in nature but not one that has to overcome itself in sublation to become self-conscious. Rather, the idea is the explosion of itself, and spirit is a bomb. As Malabou acknowledges that, "If we didn't explode at each transition, if we didn't destroy ourselves a bit, we could not live. Identity resists its own occurrence to the very extent that it forms it."[25] At the extreme, this destructive plasticity is so radical that we can no longer recognize who we are, but this is the case even when we think we do recognize ourselves and each other. Or as Malabou expresses it in *The New Wounded*, "What scorches the symbolic is the *material* destruction of the Thing."[26] And this material character of destructive plasticity identifies plasticity as other than writing.

A healing that leaves a scar or a trace works according to the model of writing because we can always read the traces that the injury shows. On the other hand, regeneration operates according to a different model, that of cloning. For Malabou, "when a lizard's tail grows back, it leaves no trace of the amputation at all."[27] This finite reconstitution of an organ is for Malabou "a regeneration of difference."[28] What she calls the paradigm of the salamander permanently erases writing by means of this replication, which is a change of difference. Difference changes, it changes form, and it does so without leaving any trace. The change in form that leaves no trace is also for Malabou a response, and to the extent that we recognize it becomes our responsibility. This is the responsibility that she invokes in her question and title "*What Should We Do with Our Brain?*"

In her essay, Malabou says that plasticity is "*the resistance of différance to its graphic reduction.*"[29] Writing is a non-present / non-absent absence, where traces manifest what they can never fully present. But according to Malabou, Derrida himself later abandoned his efforts to deconstruct presence, primarily with the shift in focus to the "undeconstructible." The undeconstructible, "that Derrida outlined in his late work under the names of 'justice' or 'democracy,'" is a name for something that comes back and regenerates like a salamander.[30] "The un-deconstructible is not of the order of presence," Malabou claims, "but it is just as much a form of resistance to the text."[31] This means that Derrida himself grappled with the metamorphosis of writing in the name of the undeconstructible, and we have translated this change into religious, ethical, and political categories. Malabou's philosophy helps us to understand how and why Derrida's philosophy changes, even if it does not simply turn. And she helps us read deconstruction as a form of materialism.

For Malabou, the brain is the locus of the self, as well as the place where history, biology, and politics happen. It is also where deconstruction happens, if there is such a thing. She charges Derrida with failing to thoroughly think deconstruction's implication in the sciences, including neurology and biology. Based on her work on plasticity, she thinks that "the time has come to elaborate a new materialism, which would determine a new position of Continental philosophy vis-à-vis the humanities and biological sciences."[32] What is this new materialism?

New Materialism (sometimes called neo-materialism by Rosi Braidotti) is a name that emerges in the 1990s, centered around interpretations of Merleau-Ponty, Deleuze and Irigaray by theorists such as Braidotti, Jane Bennett, William Connolly, Manuel Delanda, and Isabelle Stengers. The New Materialism offers resources to think about materialism otherwise than as a reductionist and determinist atomistic materialism, in concert with systems theory, chaos theory, and complexity theory. Here, being is not reduced to its smallest components or building blocks, but it is always in dynamic transformation. Malabou picks up on the phrase "new materialism" as a way to situate her work on plasticity, despite her focus on Hegel, Heidegger, and Derrida.

Influenced by Sartre and Bergson, Derrida resists the philosophy of materialism, seeing it as a simple deterministic and mechanistic theory of the world. His early work on Husserl showed the aporias that Husserl's philosophy kept coming up against as he tried to steer between a transcendental logicism and a phenomenological empiricism. These aporias affect not only philosophy and epistemology but also science and mathematics.[33] As his philosophy developed, Derrida was less and less explicitly engaged with the natural sciences, and Malabou has shifted to paradigms of neurology and biology to offer a corrective to this limit of Derrida's philosophy.

What contemporary research in brain sciences shows is how this divide between mechanism and spirit comes undone. The brain is fully material and it is fully spiritual at the same time, provided we understand spirit in non-teleological terms. Malabou states that "we persist in thinking of the brain as a centralized, rigidified, mechanical organization, and of the mechanical itself as a brain reduced to the work of calculation."[34] But this understanding is precisely what plasticity undermines because "plasticity perhaps designates nothing but the eventlike dimension of the mechanical."[35] Plasticity allows us to see the event in the mechanism, the "spirit" in the material, without it thereby ceasing to be material.

For Malabou, plasticity "is able to *momentarily characterize the material organization of thought and being*," which is why "we should certainly be engaging deconstruction in a *new materialism*."[36] The new materialism is a biological materialism of form, the plasticity of form. In an essay on Darwin and natural selection, Malabou affirms the plasticity of biological evolution: "Indeed, plasticity situates itself effectively at the heart of the theory of evolution."[37] Natural selection reveals the plasticity of the organization and structure of the organism at the level of both species and individual. The organism's variability indicates a process of transformation and selection, and this system that evolution constitutes "hinges on plasticity understood as the flexibility and fluidity of structures on the one hand and plasticity understood as a natural decision of viable, durable forms likely to constitute a legacy or lineage."[38] Natural selection is not teleological, but it works because of its inherent plasticity.

Malabou focuses on the key role of plasticity in Darwinian evolution, which is affirmed by Darwin himself. She argues that we need a social understanding of selection that is closer to this biological model, which is provided by Deleuze's interpretation of Nietzsche's eternal return. Malabou says that, for Deleuze, "selection is a return, but a return that is not the same. It is productive repetition of difference, we read in *Difference and Repetition*; and the eternal return signifies that being is selection."[39] Deleuze's idea of selection works in the same way that biological selection does—in a plastic manner. Malabou also appeals to contemporary neurology for a biological understanding that implicitly incorporates this logic of social selection along with natural selection. She focuses on the theories of Jean-Pierre Changeux, Philippe Courrége, and Antoine Danchin who develop a "mental Darwinism," which constitutes a form of "epigenesis by the selective stabilization of synapses."[40] Genetics comprises the data for cells and organisms, while epigenetic modifications introduce "variability that depends for an essential part on environmental influence, on education, and on experience which Darwin greatly helps us to think."[41]

Epigenetics is a new frontier in biological evolution and neurology. Epigenetics does not replace genetics, but it shows how specific actualizations and modifications of genes occur.[42] The genetic data comprises the envelope within which selection occurs, but epigenetics concerns the actualization of this data, or the selection of a particular history. Malabou's recent work shows how plasticity is "the epigenetic variable *par excellence*" that lies "at the heart of the relationship between variation and selection." The environment triggers certain histones in an epigenetic manner, and these hormones shape the individual organism in a way that is inheritable, despite the neo-Darwinian dogma. The inheritance of acquired characteristics sounds Lamarckian, but this neo-Lamarckianism of epigenetics offers a new way to think the link between nature and culture. The philosophical understanding of the selection of cultural forms based on Deleuze's reading of Nietzsche operates based on "a principle of variation-selection analogous to the one that operates in nature."[43] A sophisticated understanding of Darwinian evolution, the newer science of epigenetics, contemporary neurology, and Continental philosophy all converge on the plasticity of organic and cultural form.

Finally, in an essay on "The Future of Derrida," Malabou shows how her understanding of epigenetics and biological materialism can be a way to refashion rather than repudiate Derrida's philosophy. Here Malabou opposes the themes of messianicity and the undeconstructible that characterize Derrida's later work because they undermine an operative notion of time. She claims that in works such as "Faith and Knowledge," for example, "time as such has been dissolved into messianity."[44] This dissolution of time undoes the future except as a shadowy challenge to any concrete anticipation of it. "Exploring the neurological concept of plasticity," however, is for Malabou "a way to look for a new systematic question of time opposing messianity."[45] Plasticity enters into the materiality of the biological system and offers a new form of temporality that opposes the temporality of deconstruction and engages anew the "dialogue between determinism and freedom."[46]

This dialogue operates at the level of the interaction between genetics and epigenetics, as we have seen. Epigenetics opens up a space for the role of education and culture to shape an organism in powerful ways. Malabou links epigenetics to a reading of Kant, who speaks in the *Critique of Pure Reason* of the "epigenesis" of reason. Epigenesis refers to "a biological theory that opposes preformation," where the individual begins with an unformed material in which "the form emerges gradually, over time."[47] Kant cites the epigenesis of reason, but then limits it because he cannot imagine that *a priori* concepts or categories could evolve. For Malabou, this opens a space for the radicalization of Kantian epigenesis: "If reason is creative and self-formative, we are then allowed to say that the transcendental itself is plastic, and that there must be a kind of experience within the realm of the a priori."[48] She admits that Kant would not have accepted the notion of a plasticity of the transcendental, but that Hegel would and does.

At the end of her essay, Malabou contests the idea of the undeconstructible. She says that "it arbitrarily both limits deconstruction and marks its sovereignty."[49] On the contrary, Malabou confesses her faith that nothing is undeconstructible. The plasticity of the transcendental means that everything, including the transcendental, is deconstructible. And this opens a genuine future, with and against Derrida. We might want to cling to Derrida and resist this understanding of plasticity, time and future. Or we might want to view deconstruction itself as plastic, which is less a betrayal of Derrida than a way of opening his philosophy to a future, which is not his own.

We know that Derrida himself tried to anticipate and affirm the future, even though he knew that he could not anticipate his future, or that of deconstruction. There is no absolute necessity that the future of Derrida be Malabou's, but there is no a priori reason that it could not. I have tried in this chapter to read Malabou's challenge to deconstruction in a way that is compatible in some ways with Caputo's affirmation of the religious passion of Derrida's philosophy, even though it reaches a limit in technical terms. At the same time, I do think that there is a perspective in which we can view Derrida's thought genetically outside of an envelope of writing, and that this can still be faithful, at least to the spirit of deconstruction. Derrida enlarges our understanding of writing; he generalizes it. But he becomes more difficult to read and to appreciate his importance as this paradigm of writing or arche-writing recedes in significance. It's not a simple, straightforward linear replacement of one motor scheme by another, but a complex interaction that does occur in time.

Malabou's profound philosophical work challenges deconstruction. At the same time, there is still a relation to deconstruction, as she herself constantly affirms. She is both a critical reader of Derrida as well as a powerful philosopher in her own right. And she helps us understand the ways in which the natural sciences are crucial for what is called Continental philosophy and vice versa. This connection is vital not only for the future of deconstruction, but for humanity as well, given the ecological situation of our time, including the limits of economic growth given finite natural resources. We need a philosophical and theological ecology in a broad sense, and Derrida and Malabou, as well as Deleuze and Guattari and more generally what is called New Materialism, provide theoretical resources. In the final chapter, I delve into some philosophical aspects of theoretical quantum physics, and see how a contemporary philosopher of science, Karen Barad, uses Derrida's philosophy to make sense of reality at the subatomic level.

8

Quantum Derrida
Barad's Hauntological Materialism

Faith is a cascade.

—**Alice Fulton, "Cascade Experiment"**

At the end of *The New Wounded*, Malabou suggests that a wounded subject could be so intensely wounded that she could no longer be capable of responding to a person or a situation with affect or transference. She asks, "How could we deny that the new wounded *call responsibility into question?*"[1] If the inability of the patient to respond calls into question responsibility, this suggests that "between psychoanalysis and neurology, it is precisely the sense of 'the other' that is displaced."[2] If the other is displaced from subjectivity in neurology and neurological brain damage, then we might have to seek it elsewhere, on the basis of what Malabou calls "a nontransferential clinic," or an analysis that does not depend primarily on the other's conscious recognition.

One of the ways I am trying to read Derrida is by displacing his philosophy to an other context, one that is more explicitly material and plastic. For Malabou, plasticity is the sign of a neurological paradigm and the site of a biological materialism. My turn to the work of Karen Barad here is not meant to invalidate the significance of biology or to undermine the importance of Malabou's philosophy, but to offer another displacement, a "nontransferential clinic" of quantum physics where a quantum Derrida operates. This shift to the quantum level, and in particular quantum field theory, allows us to further reflect on two threads that have already been acknowledged in this book.

The first theme is the underdeveloped third entity of Derrida's questioning of Heidegger's threefold schema of stone, animal and human. It appears that in his late work Derrida was obsessed with these Heideggerian theses from Heidegger's lecture course on *The Fundamental Concepts of Metaphysics*. Derrida intervenes into Heidegger's theses most intensively on the notion of the animal. Heidegger distinguishes the animal, which is poor in world, from the stone, which simply lacks a world, and the human being, who possesses the capacity of world-building. Derrida challenges the separation Heidegger sets forth between the animal and the human. In Chapter 6 of *Of Spirit*, for example, Derrida challenges Heidegger's denial of world and spirit to the animal in his infamous Rectorship address. Here, Heidegger claims that "the world is always a spiritual world," and "the animal has no world"; therefore "the animal has no spirit since, as we have just read, every world is spiritual. Animality is not *of spirit*."[3]

Derrida does not really reflect on the situation of the stone in his multiple engagements with Heidegger, although he does express astonishment that Heidegger uses a particular example here rather than a general category of an inanimate object. He asks: "Why does he take the example of an inanimate thing, why a stone and not a plank or a piece of iron, or water or fire?"[4] At the same time, as we have already seen, Derrida says in "Rams" that "for reasons I cannot develop here, nothing appears to me more problematic than these three theses."[5] He not only questions the thesis about the animal; he also implies that the thesis about the stone being worldless is problematic, at least in comparison to the animal and the human. A stone is an object, and objects are much more complicated than we might suppose, as discussed in Chapter 5. In this chapter, however, I want to consider what kind of objects are subatomic particles?

The second theme of Derrida's that I want to develop here in relation to quantum physics is the provocative claim that "all other is all other." This the phrase suggested in the last chapter of *The Gift of Death*, the idea that "*tout autre est tout autre.*" In *The Gift of Death*, Derrida reads Kierkegaard's *Fear and Trembling*, with its reflection on the Aqedah where Abraham nearly sacrifices Isaac. In his reading, Derrida generalizes the situation of sacrifice treated in the biblical story, and he understands that the ram substitutes as a sacrifice for Isaac but that does not do away with the sacrificial economy with which we continue to operate. Our civilized society "*puts to* death or . . . *allows* to die of hunger and disease tens of millions of children . . . without any moral or legal tribunal ever being considered competent to judge such a sacrifice, the sacrifice of others to avoid being sacrificed oneself."[6] He says that we possess no authority, institution, or criteria to decide or "to determine with any degree of certainty who is responsible or guilty for the hundreds of thousands of victims who are sacrificed for what or whom one knows not, countless victims, each of whose singularity becomes each time infinitely singular, every other (one) being every (bit) other."[7]

Here in *The Gift of Death*, the phrase *tout autre est tout autre* is translated as "every other (one) being is every (bit) other." Every other individual person is absolutely unique and other, as valuable and worthy in his or her singularity that brooks no logic of comparison, even though we carelessly compare all the time. Derrida then, following Kierkegaard in a way, extends this phrase to indicate divinity: "'Every other (one) is God,' or 'God is every (bit) other.'"[8] If God is the wholly Other, the source of absolute alterity, then it makes sense to think of God as the Other. But again, Derrida generalizes: he extends this divine alterity to every one as other. He undermines the separation that Kierkegaard upholds between humanity and divinity. Here God and the person are substitutable in their infinite otherness.

In *Fear and Trembling*, Kierkegaard, writing as Johannes de Silentio, posits a religious realm that lies beyond that of ethics. Ethics for Silentio is the home of the rational universal, which means that religion becomes, strictly speaking, absurd. Abraham is the paradigm of religious faith because he obeys the command to kill his son even though it is horrific and immoral. Abraham has faith that he will retain Isaac and the promise of his descendants, despite his expectation that he will kill Isaac. Derrida generalizes the situation of Abraham's faith, to indicate the sacrificial situation that confronts us all, whether we believe it or not.

At the end of the book, Derrida shifts from Kierkegaard to Nietzsche. In the *Genealogy of Morals*, Nietzsche meditates on the phrase from the Gospel of Matthew (6:19–21) that admonishes Christians not to store their treasures here on earth, but to "lay up for yourselves treasures in heaven," "for where treasure is, there your heart will be also."[9] Derrida follows Nietzsche in his critique of Christianity as the spiritualization of an economic exchange, and we saw some of the effects of this economy in Chapter 2, as reflected in *Merchant of Venice*. The earthly treasure is subordinated to a heavenly treasure that is infinitely more rewarding because it is deferred. Similarly, Christ's teaching, reflected in the Sermon on the Mount, introduces an infinite asymmetry, a "gift, a love *without reserve*" that reproduces the economy of sacrifice at a higher level.[10] This is what Nietzsche refers to as the "stroke of genius called Christianity": "In questioning a certain concept of repression that moralizes the mechanism of debt in moral duty and bad conscience, in conscience as guilt, one might develop further the hyperbolization of such a repression."[11] The logic of sacrificial debt is not abolished in Christian sacrifice; it is universalized and infinitized. Jesus's sacrifice does not put an end to sacrifice; it interiorizes sacrifice as self-sacrifice of one's desire.

At the end of *The Gift of Death*, Derrida argues that "if there is such a thing as this 'stroke of genius,' it only comes about at the instant of the infinite sharing of the secret."[12] What is the secret of Christianity? It is the secret of sacrifice as debt and, more important, as credit. Christians are those who believe in this hyperlogic of sacrifice, this unbelievable cosmic expansion of debt that exposes us all to sacrificial death unless we can be redeemed by our belief. Here is

> the reversal and infinitization that confers on God, on the other or on the name of God, the responsibility for that which remains more secret than ever, the irreducible experience of belief, between credit and faith, the *believing* suspended between the credit of the creditor and the credence of the believer. How can one *believe* this history of *credence* or *credit*?[13]

There is a knot of belief, of credit and debt, at the heart of Christianity. But it applies not only to Christianity because this Christian logic becomes the logic of the West, what Derrida calls "globalatinization" in his essay "Faith and Knowledge."

What about science? Does modern science, which arose in what we call the "West," contain this structure of belief and sacrifice? Do we believe in the existence of subatomic particles such as quarks, for example? What credit is extended to scientific procedures, theories, and facts, and how do these practices of credence or skepticism place us in debt? A new discipline of science studies has emerged at the end of the twentieth century to reflect not only on the history and politics of science, but also a sociology of science practice. Probably the most influential book about science on intellectual academic non-scientists in the twentieth century is Thomas Kuhn's *The Structure of Scientific Revolutions*. Kuhn pays attention to the history of science and distinguishes between normal science according to an established paradigm, and revolutionary science, whereby one paradigm is replaced by another. In his usage of the term "paradigm," Kuhn draws on discussions with Stanley Cavell and others about Wittgenstein, and he makes an explicit analogy between scientific and political revolutions. Furthermore, Kuhn suggests that the replacement of one paradigm by another is not due to logic, but operates more like a conversion. In this situation, because rules are paradigm determined, a scientist cannot rely on the evidence "provided by problem-solving." Ultimately, the scientist must "have faith that the new paradigm will succeed with the many large problems that confront it, knowing only that the older paradigm has failed with a few. A decision of that kind can only be made on *faith*."[14]

In Derridean terms, in what ways is science sacrificial? In the third essay of his *Genealogy of Morals*, Nietzsche argues that science remains under the sway of an ascetic ideal, an unquestioned allegiance to the will to truth. Asceticism is the sacrifice of earthly pleasures for spiritual rewards. Our "*modern science*," Nietzsche declares, "is the best ally the ascetic ideal has at present, and precisely because it is the most unconscious, involuntary, hidden, and subterranean ally!"[15] Does contemporary science remain the best ally of the ascetic ideal, or does it represent something different?

I am not a scientist and lack the expertise in experimental and mathematical methods to evaluate the *scientific* legitimacy of science. Part of the reason for this is the hyperspecialization of knowledge in modern and contemporary civilization, which sunders scientific technical knowledge from more theoretical humanities-based knowledge. In his later work, Derrida kept a critical distance from the natural sciences, but recent expressions of Continental philosophy have re-engaged with natural sciences, including biology, neurology, chemistry, physics, cosmology, and mathematics. In this chapter, I want to think about a Derridean science, in this case quantum physics, even though I do not possess the expertise to understand or evaluate quantum physics in strictly scientific terms.

We require intermediaries, including scientists as well as philosophers and sociologists of science, to help us interpret and understand scientific laws and theories. This necessity for intermediation is not restricted to science, of course, and exists in any area of knowledge. In addition to the need for expert scientists, however, we also learn that scientists themselves require something to inter-mediate for them. Scientific discoveries require and rely upon an apparatus. So again, what does it mean to believe in and be indebted to an apparatus, especially in areas such as particle physics where nobody is capable of seeing without the aid of complex technical apparatuses that function as prostheses?

One way to read Karen Barad's work is to see her as doing something analogous to what Derrida is doing with Kierkegaard in *The Gift of Death*: By focusing on the apparatus, she is generalizing this situation of credit and debt in quantum physics to science in general. The apparatus is the locus of belief for science, which is also the place where acesis and debt, infinite asymmetry and sacrifice, take place, even if she does not use this language. I will unpack this analogy further between what Barad is doing with quantum physics and what Derrida is doing with philosophy, ethics, and religion. At a certain point, however, the analogy breaks down in two directions: first, the situation in experimental scientific practice is absolutely different from that of philosophical reflection; and second and more important, the analogy is not just an analogy, especially if *tout autre* is *tout autre*.

In the early twenty-first century, Barad's book *Meeting the Universe Halfway* has had a significant impact on philosophers and other nonscientist academics. Barad draws on science studies, feminism, poststructuralism, and her own expertise as a theoretical physicist to fashion a powerful articulation of how the quantum world impinges on our own. According to Barad, "quantum mechanics is not a theory that applies only to small objects; rather, quantum mechanics is thought to be the correct theory of nature that applies at all scales."[16] The problem is how to understand quantum physics because the experimental results of quantum phenomena have outstripped our capacity to make sense of them.

Barad offers a new ontology, an "agential realist ontology," to account for quantum phenomena. Entities in the world not only interact; they intra-act, and the world is constituted in and through their intra-actions. "The world," Barad states, "is an open process of mattering through which mattering itself acquires meaning and form through the realization of different agential possibilities."[17] Objects are not static things, but dynamic processes that change in their interaction and intra-action with other objects. Dynamics is not what happens between things, but how these things become what they are as they transform themselves and their objects in a mutual asymmetrical process of materialization.

Barad shifts our focus from things or objects to apparatuses. Apparatuses for Barad *"are not mere observing instruments, but boundary-drawing practices—specific material (re)configurings of the world—which come to matter."*[18] Here apparatuses are not simply devices in a laboratory but "material-discursive practices" that constitute phenomena, including how we conceive their relationships and intra-activity, and ultimately our entire world of "nature" and "culture."[19] Barad is a posthumanist, challenging the separation between human culture and the objective natural world. Agency and intelligibility are not restricted to human activity, but shared among various phenomena. Phenomena are constructed, but not solely by human means.

Phenomena are "ontologically primitive relations." The understanding of these relations, or intra-actions, challenges our traditional conception of causality, and leads Barad to the claim that *"phenomena are the ontological inseparability of intra-acting 'agencies.' That is, phenomena are ontological entanglements."*[20] An apparatus is a construct that intervenes into phenomena, producing an "agential cut" that effects a separation between what we call a subject and what we consider an object. As McKenzie Wark explains, "There is no good way of discriminating between the apparatus and its object. No inherent subject/object distinction exists. There is an apparatus-phenomena-observer *situation*."[21] The apparatus is the key concept here, loosed from the framework of traditional scientific boundaries.

In *Meeting the Universe Halfway*, Barad offers a reinterpretation of Neils Bohr that counters how he is sometimes downplayed in comparison to Werner Heisenberg and Erwin Schrödinger. In doing so, she revisits the famous double-slit experiment, where a stream of electrons is directed one by one at a barrier with two openings, with a screen behind it to record the location that the electrons strike the screen. The issue here is how the experiment presumes a number of discrete electrons, but the resulting pattern is an interference pattern, which is the result of wavelike behavior. "Unlike the behavior of water waves, which go through both slits at once, the electrons are sent through one at a time. Does an *individual* electron 'interfere' with itself? Does a *single* electron somehow go through both slits at once? How can this be?"[22] So is an electron a wave or a particle?

Bohr relies on a variation of this experiment to formulate his notion of complementarity. He conducts a thought experiment that suggests that if a device could measure which path the electron takes—which slit it goes through—then the interference pattern would disappear because the result would look like the electrons are simply particles. This is part of a theoretical dispute about the nature of quantum phenomena with Einstein. Bohr claims that for subatomic quantum particles, "wave and particle behaviors are exhibited under *complementary*—that is, *mutually exclusive*—circumstances."[23] Bohr is the originator of the so-called Copenhagen interpretation of quantum mechanics, which is the idea that the strange behavior of these particles is not due to our inability to observe or measure them correctly, but somehow intrinsic to the phenomena themselves.

Barad argues that complementarity depends on the experiment, and more importantly, the apparatus. If the electron double slit experiment could be conducted with a measurement that determines which path each electron passes through, then the result is a classical one that makes it look like the electron is a particle. But when the experiment is conducted as it usually is, without determining which slit each electron passes through, the electrons demonstrate the interference pattern that is common to wave interaction and superposition. She cites later experiments of Scully et. al. in the late 1980s and early 1990s that are able to determine which path the electron takes, and these experiments show that Bohr was correct. Wave particle complementarity exists, and this complementarity is an ontological notion. "What is the result?," Barad asks. She says that "despite the lack of disturbance [that was the counterargument of Einstein and others, that the striking of the screen 'disturbs' the experiment and creates the appearance of interference] the experimenters nonetheless confirm the existence of which-path-interference complementarity."[24] Complementarity is not the result of an epistemic uncertainty; it is the result of an ontological indeterminacy.

Many nontechnical readers of quantum physics cite the famous Heisenberg "Uncertainty Principle," which states that the measurement of a particle's location cannot be fixed if we want to specify its momentum, or vice versa. This uncertainty exists across a very tiny range, which is measured by a fraction of Planck's constant, h. Heisenberg's uncertainty is often seen as an epistemological uncertainty, due to the limitations of our measuring devices. But Bohr argued that our lack of knowledge is ontological, based on a fundamental indeterminacy of reality. Barad says that

> Bohr understands entanglements in ontological terms (what are entangled are the "components" of phenomena. For Bohr, phenomena—entanglements of objects and agencies of observation—constitute physical reality; phenomena (not independent objects) are the objective referent of measured properties. *Complementarity is an ontic (not merely an epistemic) principle.*[25]

Phenomena are entangled, and their intra-actions can be teased out by interacting (and intra-acting) with them by means of an apparatus.

What does this understanding of quantum physics have to do with Derrida? Although Barad does not explicitly cite Derrida in *Meeting the Universe Halfway*, she does engage with Derrida's work in other articles. I will look at two of them to show how Barad understands Derrida's philosophy as a hauntological materialism.

In "Quantum Entanglements and Hauntological Relations of Inheritance: Dis/continuities, SpaceTime Enfoldings, and Justice-to-Come," an article published in the journal *Derrida Today* in 2010, Barad uses Derrida's term "hauntology" from *Specters of Marx* to describe the world of quantum phenomena. She acknowledges the materialist readings of Derrida by Vicki Kirby and Astrid Schrader as influences for her interpretation. In this article, Barad revisits some of the quantum material from *Meeting the Universe Halfway*, including her understanding of Bohrian indeterminacy. She also reflects on how quantum superposition indicates a state of quantum entanglement.

The most famous example of this superposed state is Schrödinger's hypothesis about a cat whose life or death is determined by the state of decay of an atom. If the atom decays, it releases a gas that kills the cat. The classical presumption is that the atom is either in a state of decay or not, and therefore the cat is either alive or dead. The quantum suggestion is that somehow the cat is in a strange intermediate state, neither alive nor dead, or else both alive and dead at the same time.

Barad clarifies this situation. She argues that:

> A *quantum superposition* is a nonclassical relation among different possibilities. In this case, the superposition of "alive" and "dead" entails the following: it is not the case that the cat is either alive or dead and that we simply do not know which; nor that the cat is both alive and dead simultaneously (this possibility is logically excluded since "alive" and "dead" are understood to be mutually exclusive states); nor that the cat is partly alive and partly dead (presumably "dead" and "alive" are understood to be all or nothing states of affair); nor that the cat is in a definitive state of being not alive and not dead (in which case it presumably wouldn't qualify as a (once) living being). Quantum superpositions radically undo classical notions of identity and being (which ground the various incorrect interpretative options just considered). Quantum superpositions (at least on Bohr's account) tell us that being/becoming is an indeterminate matter: there simply *is not a determinate fact of the matter* concerning the cat's state of being alive or dead. It is a ghostly matter! But the really spooky issue is what happens to a quantum superposition when a measurement is made and we find the cat definitively alive or dead, one or the other.[26]

Quantum superposition means that phenomena are entangled in such a spooky way that there is no simple either/or.

Superposition is related to entanglement, which is the entanglement of particles and states in a nonclassical way. These states have to be measured, which makes the entanglement of a live and dead cat "decohere" into a classical situation where the cat is either dead or alive. Furthermore, this situation of quantum entanglement suggests nonlocality, which Einstein derisively called "spooky action at a distance." Given two entangled states or particles, say two electrons whose spin is correlated, then measuring one of the electrons gives the spin of the other one, no matter how far apart they are. And the result of this measurement happens *faster than the speed of light*. This understanding of entanglement is the result of Bell's Theorem, which John Bell proposed in the 1964 as a way of answering the challenge that Einstein proposed along with Boris Podolsky and Nathan Rosen known as the EPR paradox. Alain Aspect and others then carried out experiments in the early 1980s that confirmed Bell's theory and the predictions of quantum mechanics against Einstein's criticisms.

According to Barad,

> *Quantum entanglements* are generalised quantum superpositions, more than one, no more than one, impossible to count. They are far more ghostly than the colloquial sense of 'entanglement' suggests. *Quantum entanglements* are not the intertwining of two (or more) states/entities/events, but a calling into question of the very nature of two-ness, and ultimately of one-ness as well.[27]

She argues that Derrida's philosophy offers a better way to understand this situation than many of the interpretations supplied by quantum physicists. Words, concepts, phenomena are entangled in complex ways, and deconstruction attends to the manner in which such phenomena are spookily entangled.

This is a materialism of a sort, but a very strange kind of materialism, that Barad calls a hauntological materialism. Our entangled intra-actions as large slow beings repeat in a different way the relations among subatomic particles. We come to face our past lives and previous historical figures as ghosts, exerting a hauntological influence on the present. We encounter these ghosts of our past in a time "out of joint," as Derrida quotes Hamlet in *Specters of Marx*. Furthermore, these ghosts are

> encountered in the flesh, as iterative materialisations, contingent and specific (agential) reconfigurings of spacetimematterings, spectral (re)workings without the presumption of erasure, the 'past' repeatedly reconfigured not in the name of setting things right once and for all (what possible calculation could give us that?), but in *the continual reopening and unsettling of what might yet be, of what was, and what comes to be.*[28]

For Barad, there is an ethics of responsibility derived from Derrida's work that offers a sense of justice or doing justice in relation to her understanding of quantum physics.

There is an ongoing materialization of matter, and an ongoing dynamic production of space and time, as the framework within which phenomena interact and intra-act. The apparatus names the agential cut that gives us the appearance of disparate phenomena, but we are still haunted, affected, and effected by the entangled superposition these phenomena remain, at least virtually. Responsibility involves "facing the ghosts, in their materiality, and acknowledging injustice without the empty promise of complete repair."[29] Quantum phenomena are intrinsically spooky, and they haunt us no less than larger scale phenomena.

In her article on "Quantum Entanglements and Hauntological Relations of Inheritance," Barad does not simply comment on Derrida's work. She quotes him at points, and then shows how his ideas are relevant to the world of quantum physics. In another article, "On Touching: The Inhuman That I Therefore Am," she reflects on the paradox of touch within the context of quantum field theory, and connects this idea to Derrida's book *On Touching—Jean-Luc Nancy*. The paradox Barad starts with is that for physicists, phenomena do not actually touch. In physical terms, touch is an electromagnetic interaction: "there is no actual contact involved."[30] Touch is a paradoxical phenomenon, even at the most basic physical level. Things get even more strange when Barad discusses quantum field theory.

According to Barad, in quantum field theory "there is a radical deconstruction of identity and of the equation of matter with essence in ways that transcend even the profound un/doings of (nonrelativistic) quantum mechanics."[31] In quantum field theory, a particle is the expression of the entire field at a specific point, a quantum. And conversely, the field is the expansion or generalization of the particle. For example, the photon is the quantum of an electromagnetic field. The particle and the field are two complementary ways of understanding the same phenomenon. Furthermore, the particle and the field are entangled with a void. Particles do not exist within a void; they are constitutively entangled with the void. The void is not a simple vacuum; it is "a living, breathing indeterminacy of non/being."[32] The vacuum is full of virtual particles.

What is a virtual particle? When quantum physicists explore the ways that particles act, they discover that these particles give rise to virtual particles, particles that appear to jump into existence and then just as quickly jump back out of it. There is a sort of perversity intrinsic to these fundamental subatomic particles. Richard Feynman said the following about electrons: "Instead of going directly from one point to another, the electron goes along for a while and suddenly emits a photon; then (horrors!) it absorbs its own photon. Perhaps there's something 'immoral' about that, but the electron does it!"[33]

Barad claims that "*virtual particles are quantized indeterminacies-in-action.*"[34] These particles exhibit a propensity to touch and therefore self-touch every other possible particle as part of their paradoxical perversity. In order to handle the queer infinities produced by virtual particles and their interactions, physicists have to renormalize these infinities in practice. But that does not eliminate the strangeness of these virtual particles; it merely allows us to deal with them.

For Barad, quantum touching is a strange phenomenon, where every other particle intra-acts with itself and every other particle. Although she does not specifically refer to Derrida's phrase from *The Gift of Death*, *tout autre est tout autre*, I think it fits even better her description of the strange interactions and intra-actions of quantum field theory. These intra-actions can be characterized as what Barad calls an "infinite alterity," and she summons a Derridean responsibility to do justice to this insensible indeterminacy of being.[35] In a quantum field where quantized particles self-touch in virtual ways, every other is indeed every other. Self-touching is a kind of self-sacrifice because these virtual particles are created and destroyed in a wink of existence. Matter is transient: "No longer suspended in eternity, matter is born, lives, and dies."[36] Here is a certain end of the Heideggerian thesis, where the inanimate thing is not a worldless stone but a virtual particle that inhabits *worlds without end*.[37] Barad refuses the Heideggerian refusal of a world for the stone; she offers an opening to the inhuman that we also are. For Barad, ethics involves "a recognition that *it may well be the inhuman, the insensible, the irrational, the unfathomable, and the incalculable that will help us face the depths of what responsibility entails*."[38] Responsibility, sacrifice, and the gift of death inhabit and haunt quantum worlds, even when the world as such is gone. There are no worlds, only superposed islands of indeterminacy. Waves, particles, void, all entangled.

Karen Barad is not the only philosopher who is thinking about quantum physics. In recent years, François Laruelle has developed his non-philosophy into what he calls a non-standard philosophy based on quantum phenomena. In his book *Philosophie Non-Standard*, published in French in 2010 and not yet translated into English, Laruelle makes explicit use of concepts from quantum physics to explicate non-philosophy. He argues for a more quantum-theoretical conception of the Real, and suggests that the wave-particle complementarity of quantum physics offers a better model for theoretical thinking than atomic entities. Laruelle says that non-philosophy does not intervene directly in quantum physics, but rather takes the same relation to philosophy as quantum physics does to classical physics, which is less one of inclusion and more one of generalization along reconstituted lines.[39]

Laruelle claims that while traditional philosophy deals with discrete, corpuscular objects and concepts, these concepts are deformed by the implications of wave superposition. Laruelle wants to superpose philosophical concepts together, which is a scientific operation, or what he calls "the unity of a fusion of science and philosophy under science."[40] This fusion is the result of superposition, which constitutes a radical immanence, not absolute in the sense of being opposed to transcendence, but radical in the sense of being composed immanently by super-imposing concepts and seeing how they amplify or reduce each other. Non-standard philosophy works with superposed terms rather than simply discrete classical ideas. These are complex ideas as opposed to simple ideas, in the way that complex numbers include not only real numbers but also an imaginary component, which is the square root of -1, written as i. Laruelle says that an imaginary number can be thought of as a "quarter-turn" backwards from a circle, which is an immanent subtraction from the binary doublet of immanence/transcendence.[41]

Laruelle's terminology and writing is incredibly complex, and it is made more so by the fusing together of philosophical ideas and concepts, which he subjects to his own rigorous non-philosophical formulations, with quantum ones. I am not going to try to fully explain Laruelle's philosophy—and in truth I could not do it justice—but I want to give a sense of what he is doing and how it compares to Barad's work. Barad is using the language of Derrida, Levinas, and other poststructuralist philosophers to articulate and explain quantum phenomena, and this helps us read these theorists beyond any sort of naïve humanism or linguisticism. Laruelle thinks that poststructuralist philosophers, despite their desire to expose philosophy to an infinite alterity, end up confirming and upholding the traditional status of philosophy and its authority to interpret the world. For Laruelle, non-philosophy is the attempt to use philosophical ideas and categories without reinforcing the sufficiency of philosophy's mastery of the world. Laruelle uses quantum terminology and concepts to rethink and refashion what normally goes by the name of philosophy.

Derrida himself claims that what Laruelle does is to reinstate a kind of authority of science over philosophy. In a debate from 1988, Derrida suggests:

> Then you went on to oppose to this description [of the sufficiency of philosophy] this new science, which you distinguished from its political, social, etc., appropriations, and there, obviously, I had the impression that you were reintroducing philosophemes—the transcendental being only one of them—into this description, this conception of the new science, the One, the real, etc. There, all of a sudden, I said to myself: he's trying to pull the trick of the transcendental on us again, the trick of auto-foundation, auto-legitimation, at the very moment when he claims to be making a radical break.[42]

And Laruelle replies by acknowledging that from a philosophical standpoint what he's doing appears crude and contradictory, but that he allows himself "the right, the legitimate right, to use philosophical vocabulary non-philosophically." Laruelle claims that although Derrida uses philosophy against itself, against its own pretensions to absolute knowledge, he still does so in the name of something like philosophical logic itself. Laruelle appeals to science, which "makes a non-positional, non-thetic use of language" to think outside of and beyond philosophy itself.[43] And in *Philosophie Non-Standard*, Laruelle appeals to the science of quantum physics to think philosophy otherwise.

Gilles Deleuze is impressed by Laruelle's project, although he also suggests that Laruelle grants too much authority to science. In his last book, *What is Philosophy?*, published in French in 1991 and co-authored with Félix Guattari, Deleuze writes in a footnote that "François Laruelle is engaged in one of the most interesting undertakings of contemporary philosophy. He invokes a 'One-All' that he qualifies as 'nonphilosophical' and, oddly, as 'scientific,' on which the 'philosophical decision' takes root. This One-All seems to be close to Spinoza."[44] Deleuze appreciates the effort to construct a non-philosophy along Spinozist lines, but he hesitates before Laruelle's evident scientism, commenting that "we do not see why this real of science is not nonscience as well."[45]

Part of the question at stake is the nature and role of science, and to what extent the appeal to quantum physics, mathematics (as in the set theory invoked by Alain Badiou), or biology involves a kind of trumping of philosophical practice by the discipline of the natural sciences. Unfortunately, most scholars trained in philosophical or other humanities-based disciplines lack the technical expertise to evaluate scientific ideas, but even scientists trained in a certain discipline lack the expertise to evaluate another discipline. From my perspective, after his early work on geometry, anthropology, and linguistics, Derrida is a little too skeptical of the natural sciences, whereas sometimes philosophers, such as Laruelle, Badiou, Meillassoux, and adherents of Speculative Realism and object-oriented ontology give science too much weight. I do think, however, that we need to try to understand and think about scientific ideas, even if we are non-scientists. And this is what Barad's work gives us tools to do.

By way of a conclusion, I suggest that what Barad calls "diffraction" functions as a kind of quantum *différance*. In *Meeting the Universe Halfway*, Barad says that "there is a deep sense in which we can understand diffraction patterns—as patterns of difference that make a difference—to be the fundamental constituents that make up the world."[46] Diffraction in classical terms occurs with waves, including ocean waves. As waves interact with other waves or with an object, new patterns are formed. A diffraction pattern is also an interference pattern because of the way that waves, or waves and objects, interfere with each other. When a wave meets another wave, depending on the wavelength it can either cancel out the other wave or amplify it. We can also see an interference pattern or a diffraction pattern when a wave hits a gap or hole as in a breakwater.

Diffraction is "an entangled phenomenon" that takes place in large-scale classical phenomena, as well as in quantum subatomic phenomena.[47] Barad explains that when waves meet and interact, they combine to form a new wave. "The resultant wave," she says, "is a sum of the effects of each individual component wave; that is, it is a combination of the disturbances created by each wave individually."[48] The combination of individual waves is called superposition, a term that Laruelle also uses. Superposition works at quantum and large-scale levels, and it produces a diffraction pattern.

As McKenzie Wark explains in his discussion of Barad's work, "diffraction is not about how one thing is an imaginary reflection or double of another."[49] Difference involves a joint or co-production of new phenomena. "Diffraction is about how things pass through and produce differential patterns," and these patterns are real at the level of material-physical phenomena.[50] The apparatus is that object or entity that measures the result of this process and evaluates in what way it is a novel phenomenon.

A diffraction pattern is a pattern that is constituted by a differentiator—a situation that prompts difference to relate to difference. When we think about differences, we usually think about difference as derivative from or subordinate to identity. How can we compare anything except a prior identity from which something differs? Here, the powerful reorientation of poststructuralism as epitomized in the philosophy of Derrida and Deleuze is the contention that differences are primary, and that differences can relate to differences without depending on a pre-given identity.

Deleuze develops this notion most explicitly in *Difference and Repetition*. According to Deleuze, in order to relate difference to difference there exists what he calls a second-degree of difference, a differentiator. He says that for this second-degree of difference to operate, differences must be organized into a series. First of all, there have to be at least two series, and second, there must be a communication between these series, a "force" or intensity that "relates differences to other differences." It is the "intensive character of systems" or series that is important here, the manner by which difference is related to difference, the construction of a second-order difference. An intensive field communicates differences to differences, and produces individuations. According to Deleuze, "once communication between heterogeneous series is established, all sorts of consequences follow within the system. Something 'passes' between the borders, events explode, phenomena flash, like thunder and lightning."[51]

In order to recognize difference as difference, or to distinguish between and among differences, we would seem to need to refer to a kind of resemblance or identity. We assume a privileged path of thinking from identity to difference, but as a matter of fact it's the reverse: "Thunderbolts explode between different intensities, but they are preceded by an invisible, *dark precursor*, which determines their path in advance but in reverse, as though intagliated."[52] This is because of the pressure differential in thunderstorms: "Likewise, every system contains its dark precursor which ensures the communication of peripheral series." The dark precursor is the differentiator, and it is also what in Deleuze and Deleuze and Guattari's later work becomes the plane of immanence or plane of consistency. Deleuze explains:

> There is no doubt that *there is* an identity belonging to the precursor, and a resemblance between the series which it causes to communicate. This "there is," however, remains perfectly indeterminate. Are identity and difference here the preconditions of the functioning of the dark precursor, or are they, on the contrary, its effects?[53]

Deleuze argues here for the second viewpoint, which means that the dark precursor projects "upon itself the illusion of fictive identity, and upon the series which it relates the illusion of a retrospective resemblance." There is an identity, but this identity is a transcendental illusion cast by the shadow of the dark precursor. Resemblance and identity are illusions—"in other words, concepts of reflection which would account for our inveterate habit of thinking difference on the basis of the categories of representation."[54] The dark precursor conceals itself in its operation, and this gives rise to the illusion of identity. The dark precursor is the differentiator because, "given two heterogeneous series, two series of differences, the precursor plays the part of the differentiator of these differences."[55] The dark precursor is invisible and the path it traces only becomes visible in reverse, when the lightning strikes. These differential conditions cause phenomena to occur. Difference in itself communicates to itself through a force or intensity that Deleuze names repetition. The dark precursor is "the *disparate*," whereas resemblance is an effect or external result.[56]

A way to better understand the role of the dark precursor as the disparate is to think of it in terms of a moiré pattern. A moiré is an interference pattern that emerges when two fabrics or grids are brought together and superimposed, one on top of the other. The pattern that emerges is the difference between the two patterns. For Deleuze, differentiation occurs when two series are brought together in such a way that the differences between the two series creates a third. It's not the identity of the points in the series that creates identity here; it's the relationship of difference between the two series that generates the pattern. The dark precursor is the activity of relating series to series, without any background, and the lightning flash is what emerges when the two series are inter-related or inter-meshed. What Deleuze calls the disparate indicates the process of the production of difference. Identity is generated by and out of difference.

I think that the way Deleuze understands difference in *Difference and Repetition* is very close to how Derrida understands *différance*. Derrida coins his neologism to take account of the dynamic aspect of difference, not just the normal static comparison of how one word or object differs from another. Difference is always differentiation, and it is a dynamic process. Derrida more explicitly treats linguistics and analyzes the differential operation of language, compared to Deleuze who is less influenced by the developments of structural linguistics. The verb *différer*, Derrida explains, has two distinct meanings that converge in the word *différance*. The first involves a temporal difference: "*différer* in this sense is to temporize, to take recourse, consciously or unconsciously, in the temporal and temporizing mediation of a detour that suspends the accomplishment or fulfillment of 'desire' or 'will,' and equally effects this suspension in a mode that annuls or tempers its own effect."[57] The second meaning of the word *différer* is the more common notion of not being identical, although here Derrida also introduces a dynamism, because differences resist each other in terms of an "allergic and polemical otherness" in addition to just being different. He argues that

> the word *différence* (with an *e*) can never refer to *différer* as temporization or to *differends* as *polemos*. Thus, the word *différance* (with an *a*) is to compensate—economically—this loss of meaning, for *différance* can refer simultaneously to the entire configuration of its meanings.[58]

I think that Derrida's *différance* is very close to Deleuze's concept of differentiation, although Derrida claims that the problem with the term "differentiation" is that it "would have left open the possibility of an organic, original, and homogeneous unity that eventually would have to be divided, to receive difference as an event."[59] And this is how all too many readers understand Deleuze's philosophy, as positing an original and homogeneous unity that is then divided or differentiated. But I contend that Deleuze's philosophy is not nearly so simple, and that his entire problem in *Difference and Repetition* concerns how difference relates to difference. This is a kind of diffraction pattern, to put it in the language of Barad, and it gives us a way to view *différance* as a physical and not just linguistic process. Quantum field theory involves both the dynamic temporization of time and space (understood in terms of waves or wavelike phenomena), *and* a dynamic differend as polemical opposition to identity (of a particle in relation to other particles) because it involves the "simultaneous" creation and destruction of virtual particles, which extends to take into account all of reality.

Because Derrida most explicitly concerns himself with the limits of language, the otherness in and of language, most readers assume that *différance* is a purely linguistic phenomenon. I think this is a misunderstanding of Derrida, even though he mostly investigates and applies *différance* to what Malabou calls the motor scheme of writing. We have the ability to read Derrida otherwise, which includes reading him in a more materialist way. *There is no proper Derrida*, but there are more interesting, relevant, and compelling iterations of Derrida's thought. If we read *différance* as a diffraction pattern, we see how it operates outside of language, in the world, including the world of quantum physics.

In this specific sense, then, aided by the extraordinary work of Barad, we can constitute a quantum Derrida not as the only Derrida, but one more Derrida, since there is always more than one Derrida. This is a sort of fidelity, of love, even if it is also a queer reading, as Barad would affirm. I can imagine that Derrida would be hospitable to such an understanding, even if he would also resist and undercut it, seeing as how it would necessarily deconstruct. I can also imagine that Derrida loves us and that he is smiling, as his final words instruct. And this is *undeconstructable.*

Afterword: The Sins of the Fathers—
A Love Letter

Instead of a more conventional conclusion that summarizes the book, I am offering an afterword that reflects more personally on Derrida in relation to love, fatherhood, and mourning. These reflections, rather than building on the specific chapters, offer another scene where Derrida's philosophy continues to be relevant, in many respects beyond writing. Love is at once a political, religious, and a material force, as well as a way to indicate what matters in our lives in ethical and spiritual terms.

Here I want to reflect a little more personally on what Derrida has meant to me, particularly as a kind of substitute father figure. This issue of genealogical fatherhood is not just personal; it raises the question of patriarchal sexism in philosophy (and not just in philosophy). Derrida wrote profoundly and problematically about the question of women, and his work has inspired feminist inspiration and critique. Derrida coined the word "phallogocentrism," which means that what Lacan calls the phallus structures our understanding of language and reality. The phallus is not the penis, of course, but it substitutes for the penis in an idealized but problematic way. Derrida was highly critical of Lacan, but Lacan's influence on Derrida is undeniable. Lacan's theories about the nature of the symbolic and what he calls the "Name-of-the-Father" will inform this meditation. I will draw on some of the insights of Malabou, as well as some other important women thinkers: Noëlle Vahanian, Julia Kristeva, Luce Irigaray, Bracha Ettinger, Catherine Keller, Colleen Hartung, and Katerina Kolozova.

In psychoanalytic terms, love is always ambivalent. Here in this afterword, love is a kind of love of the father, or an idealization of the father, including an idealization of Derrida as a father-figure in philosophical terms. But this, too, is ambiguous and ambivalent. According to Lacan, there is no such thing as a sexual relationship. This is a provocative claim, and what he means is that sexual connection does not occur in an exact match, one to one. There is always a distortion, a missing of the mark, that both drives sexual desire and leads to its inevitable disappointment. Lacan says that love is more correctly associated with transference. In analysis, transference is the projection of desires and ideals onto the analyst by the patient, and this can provoke a counter-transference on the part of the analyst back onto the patient. In clinical psychoanalysis, treatment depends on the ability to successfully negotiate the transference on the part of the analyst and eventually the patient. But transference is a broader and more generalized phenomenon than the role it plays in the analytic setting.[1]

Transference exists in the classroom, for example, when the student transfers certain assumptions, desires, and expectations from his or her life onto the teacher, and this also risks counter-transference back onto the student. Every situation is always already gendered, sexualized, and racialized in its dynamics, whether or not we are conscious of these dynamics.[2] In this Afterword, I am interested in the transfer from the biological father to a substitute father figure in philosophy, mainly in terms of a male subject position that has proved normative in philosophy, culture, and psychoanalysis.

Derrida was not my first philosophical love, but he has stayed with me the longest. My first love was Nietzsche, and reading Nietzsche tore me apart, and made it impossible for me to be a systematic or analytic philosopher or theologian. From Nietzsche I turned to Michel Foucault, and the impact of his *History of Sexuality* was enormous; it was mind boggling to consider the idea that our sexual desire is the essence of who we are is only a recent historical construction rather than a timeless phenomenon. In my last semester of college, I took a special topics course on "Language, Meaning and Theory from the Bible to Derrida," and in that course we read selections from *Of Grammatology*. I didn't really understand Derrida's thought, but it was provocative and transformative, the idea that the opposition between nature and culture could deconstruct. Later, in graduate school at Syracuse University, I read Gilles Deleuze and his ideas have continued to influence and inform my thinking. Instead of understanding Deleuze in opposition to Derrida, however, I ended up seeing Deleuze as much more profoundly complementary to him.

In the film *Derrida* from 2002, directed by Kirby Dick and Amy Ziering Kofman, Derrida is asked a question about who could be his philosophical mother. He answers:

> My mother, my mother couldn't be a philosopher. A philosopher couldn't be my mother. That's a very important point. Because the figure of the philosopher is, for me, always a masculine figure. This is one of the reasons I undertook the deconstruction of philosophy. All the deconstruction of phallogocentrism is the deconstruction of what one calls philosophy which since its inception, has always been linked to a paternal figure. So, a philosopher is a Father, not a Mother. So the philosopher that would be my mother would be a post-deconstructive philosopher, that is, myself or my son. My mother as a philosopher would be my granddaughter, for example. An inheritor. A woman philosopher who would reaffirm the deconstruction. And consequently, would be a woman who thinks. Not a philosopher. I always distinguish thinking from philosophy. A thinking mother—it's what I both love and try to give birth to.

Here Derrida affirms the figure of the philosopher as a male, a father figure, although he also claims that his work is dedicated to deconstructing the association of the philosopher with the father, and opening up the possibility for a woman philosopher. A woman philosopher is a thinking woman who would be at once mother and granddaughter.

On the one hand, Derrida challenges patriarchal sexism, and the intrinsically male figure of the philosopher. On the other hand, he admits that he cannot have a mother who is a philosopher, he can only imagine a future—not even a daughter but a granddaughter—where a woman could be a philosopher and therefore a mother-philosopher. This is a kind of fantasy, even if every philosopher should desire for it to become a reality. Philosophy is a chain of father figures, who substitute for the biological father in a way that structures knowledge and reality, as Lacan has theorized in his work.

"A child is being beaten," says Freud, and this constitutes a certain primal scene. Authority and punishment stem from the role of the father, who generates anxiety and fear on the part of the child. In *Civilization and Its Discontents*, Freud argues that religion stems from "the infant's helplessness and the longing for the father," although he does not fully consider how the father helps to create and reinforce this sense of helplessness.[3] The father is the source of the internalized super-ego that makes the individual unhappy throughout most of his life because the super-ego prevents the individual from expressing and enjoying his primal drives of aggression and sexuality. My argument is that one part of the drive for philosophy results from the ambivalence toward the biological father and the work of identification with idealized substitute fathers. As the Roman philosopher Seneca declares, philosophy "allows me to choose my fathers."[4]

This is where it is helpful to consider the work of Julia Kristeva. In her groundbreaking book *Revolution in Poetic Language*, Kristeva distinguishes between the semiotic and the symbolic. The semiotic chora is the name she gives to Freudian "primary processes which displace and condense both energies and their inscription."[5] The symbolic intervenes into the semiotic chora and establishes a secondary symbolic order of language and representation that constitutes the subject. We do not have any direct access the semiotic after its replacement by the symbolic, but these semiotic processes can break the bounds of the symbolic in transgressive ways. Here the semiotic is seen "as a 'second' return of instinctual functioning within the symbolic, as a negativity introduced into the symbolic order and as the transgression of that order."[6] Art, and specifically poetic language, is an irruption of the semiotic into and beyond the symbolic. Poetic language is a kind of sublimation that Kristeva values in *Revolution of Poetic Language* and later works.

Although this is much less explicit in *Revolution in Poetic Language*, Kristeva associates the mother or maternal with the semiotic, and the father with the symbolic. She affirms both the necessity of passing from the semiotic to the symbolic, and the valuing of the negativity that emerges from the rejection of certain aspects of the symbolic when the semiotic re-emerges. In *This Incredible Need to Believe*, for instance, Kristeva argues that the infant "projects itself onto a third person with which it identifies: the loving father."[7] This projection and identification are what liberate the child from the semiotic, which is depicted in Kristeva's later work as a devouring mother who threatens to prevent the child from achieving a personal identity. I think that Kristeva both overvalues the identification with the loving father and the threat of the semiotic as devouring mother. Nevertheless, I think there is something important about the process that she articulates, which enables the perpetuation of patriarchy.

The biological father, or the father figure who is closest to the infant, takes on the qualities of the loving father even as he is the source of fear and punishment as a figure of authority. Then, when and if the biological or familial father figure is seen in more ambivalent terms during youth or adolescence, there is the attachment to other figures who serve ideally if not actually as loving fathers. This is part of the origin of philosophy, among other things. For me, the ambivalence toward my biological father was partly replaced by a strong love—a projection onto and identification with the person of Jacques Derrida. He was a good father, as well as a great philosopher, partly because he challenged phallogocentrism. At the same time, we know that he was also a seductive "ladies man," who had many affairs including a famous one with Sylviane Agacinski, who had one of his children.[8]

This substitution is not one. There are multiple father figures, and for me the identification with Derrida is connected to two other male figures who served as positive role models: Charles Winquist and John D. Caputo. It was Winquist who became my mentor at Syracuse University, even though he was at the time more critical of Derrida's philosophy. Winquist represented an ideal father figure, because he loved and supported me seemingly unconditionally. It was Winquist who taught me how to read Deleuze, and who chaired my dissertation on Kant's sublime. It was Winquist who died of liver failure in 2002 as I was desperately searching for a tenure-track academic position. During my graduate study at Syracuse I was also struck by how much easier it was for me to flourish as a male doctoral student compared to many of my female friends and colleagues because of the ways that the academy and its disciplines are structured patriarchally.

Before Winquist died, he became friends with Caputo. And in some ways, Caputo came to substitute for Winquist as an academic quasi-father, or at least a loving paternal figure. After Winquist died, I reached out to Caputo and he responded positively, helping me in many ways impossible to recount or repay. Furthermore, in 2004 Caputo replaced Winquist at Syracuse University, confirming this academic filiation and investing him in my success. Of course, it was also Caputo's books and conferences at Villanova that mediated a large part of Derrida for me, beginning in 1995 at the Roundtable discussion that became the book *Deconstruction in a Nutshell*. I drove down to Villanova with another graduate student from Syracuse, and after the roundtable I was able to meet and talk to Derrida in person, and he autographed a copy of his book *Aporias* for me. A little over a month later my car was stolen off the street in Philadelphia, and this book was in my car.

I am setting up what Deleuze calls an effect series: Derrida, Winquist, Caputo. They all played a role as idealized substitute fathers, and they all helped me tremendously in my life and career, even if Derrida did not know me personally. Love always contains an element of narcissism. According to Derrida, "there is not narcissism and non-narcissism; there are narcissisms that are more or less comprehensive, generous, open, extended."[9] Narcissism is how we appropriate the other. But if we didn't add something to the other, "the relationship to the other would be absolutely destroyed."[10] As Noëlle Vahanian, a fellow graduate student with me at Syracuse and now a significant philosopher of religion in her own right, explains, narcissism "becomes a desire to know the other, for the image of self is incapable of containing the self."[11] We cannot simply love ourselves without an other, because we are not simply ourselves. Vahanian argues that in this situation, the narcissist becomes an idolator, a rebel: "eyes wide-open, this narcissist sees more than herself in another, the other in herself. She saves herself."[12]

Narcissism as love of the father is a form of identification that stabilizes and perpetuates the sexist patriarchal order. But in Vahanian's language, which also marks a shift in gender terms, this narcissism can also be seen as rebellious, and even salvific. This is incredible and actually impossible if I am just myself, relating my same male subjectivity to another male subject as role model and philosopher. Just as my attachment to a philosophical paternal figure constitutes a kind of rebellion against my biological father, my own philosophical expression involves working out a kind of idolatry against these substitute fathers. To be faithful to the father is to betray the father. This ironically also perpetuates phallogocentrism.

I don't see Catherine Malabou as a maternal figure, but she is definitely "a woman who thinks." In her book *Changing Difference: The Feminine and the Question of Philosophy*, Malabou directly reflects on complex issues of gender and sexuality from the standpoint of her identity as a woman philosopher and her viewpoint of plasticity. She argues that "to construct one's identity is a process that can only be a development of an original biological malleability, a first transformability. If sex were not plastic, there would be no gender."[13] In an essay on "The Meaning of the 'Feminine,'" Malabou develops her own thought in relation to that of Luce Irigaray and Judith Butler. She cautiously endorses Irigaray's notion of the feminine as "the fold of the lips to one another, a withdrawal that is so easy to force open, to breach, to deflower, but which at the same time also marks the territory of the inviolable."[14] There is a sense in which the feminine for Malabou consists in the inviolable: "without the feminine, the inviolable cannot be thought."

Because of its essential fragility, the idea of the feminine as the inviolable exists within a context of its actual violability. She says that "no doubt woman will never become impenetrable, inviolable. That's why it is necessary to imagine the possibility of woman starting from the structural impossibility she experiences of not being violated, in herself and outside, everywhere."[15] But this situation leads Malabou to a problem, because when we name the inviolable as the feminine, "we run the risk of fixing this fragility, assigning it a residence and making a fetish out of it." At the same time, "if we resist it, we refuse to embody the inviolable and it becomes anything at all under the pretext of referring to anyone."[16] To name the inviolable as the feminine is to "interrupt a void in difference," whereas to refuse to name the inviolable is "to refuse to interrupt a void in difference." Both stances are equally justified and equally problematic. The specification of feminine difference risks fetishizing it, while the generalization of the inviolable beyond the feminine risks diffusing and emptying it in the context of patriarchal masculinity.

Malabou complicates the already complicated relationship between the feminine and woman. She says that the terms of this relation need to be displaced, and she refers to her analysis of the exchange between Being and beings in her book *The Heidegger Change*. "Being and being change from one into the other," she writes, "that's the plasticity of difference."[17] Being—here the feminine, and beings—in this case women—"exchange modes of being." This substitutability exceeds metaphysics, because both Being and beings change in their exchange. If "substitutability is the meaning of Being," then "transvestitism comes with difference."[18] Being is not incarnated in embodied beings, but bodies manifest Being as change even as they change Being by exchanging it.

Malabou opens up the question of the feminine to the transformation of Being and the change in difference that female beings make. This is a kind of transvestitism because the woman does not remain unchanged. She refers to a place in her Heidegger book where she and her translator, Peter Skafish, decided that the word "essence" in Heidegger's philosophy is a kind of "*going-in-drag*."[19] If gender is a *genos*, a genre or an essence, and essence is always going-in-drag, then that suggests a kind of transvestitism of Being and beings, a clothing across the heart of existence. Malabou concludes that "while the feminine or woman (we can use the terms interchangeably now), remains one of the unavoidable modes of ontological change, they themselves become passing, metabolic points of identity, which like others show the passing at the heart of gender."[20] Tracing the feminine leads us to a passing that is inscribed at the heart of gender.

There is no question then that Malabou is a woman philosopher, that she *passes* for a woman philosopher, but it is not entirely clear what either of these signifiers means. She says that "if I'm a philosopher it is at the price of a tremendous violence, the violence that philosophy constantly does to me and the violence I inflict on it in return."[21] Philosophy is figured as masculine here, as the object of a "fierce quarrel" whose outcome is "ever more uncertain and unexpected," that produces "an absolute solitude."[22] Woman's liberation is essentially tied to the liberation of all of us, in our shared and unshareable absolute solitude, even those of us who do not pass as women. If Being were not change, as Malabou asserts in her interpretation of Heidegger, there would be no possibility for liberation from patriarchy.

I definitely pass as a male philosopher of religion, and in most respects I acknowledge this identity. But this acknowledgement is not unambiguous, and it comes at a cost of recognition, for myself and for others. The narcissism of my own self-identification passes through what Vahanian calls a rebellion, a revolt against the being of my being as masculine. Because my being is also change, it is also essentially a passing or a "going in drag." This means there is a part of me that does not know who or what I am, even when I perform my masculinity or benefit from its presumptions. And this is good news.

In her book *The Matrixial Borderspace*, painter and philosopher Bracha L. Ettinger develops an interpretation of what Kristeva calls the semiotic as less divorced from the symbolic. Ettinger claims that the semiotic matrixial borderspace generates symbolic patterns in a productive way. She claims that the matrix "is a concept for a *transforming borderspace of encounter* of the *co-emerging I* and the *neither fused nor rejected uncognized non-I*."[23] For her, unlike for Kristeva, there is no need for a break with the maternal semiotic in order to accede to the paternal symbolic, and the matrixial borderspace informs all three levels of imaginary, symbolic, and real. I think that Ettinger's reinterpretation of Lacan, along with her insights into Kristeva and Deleuze, provides important resources to reconceptualize Lacanian concepts in a less antagonistic way. For Ettinger,

> The matrixial psychic space concerns shareability and severality that evade the whole subject in self-identity, endless multiplicity, collective community and organized society. The matrixial borderspace is drawn and is further drawing virtual and real traumatic and phantasmatic as well as imaginary and symbolic transgressive psychic contacts by inhabitation and erotic co-tuning in the same resonance field; vibrating space or elusive time of which each participant becomes partial by its own reattunement and attention. Affective vibrations that tremble with virtual strings, body-psyche-space-time cross-imprints uncognized memory traces accumulated in several threads transform each partial-subject into some kind of mental continuity of the psyche of another partial-subject. Each psyche is a continuity of the psyche of the other in the matrixial borderspace.[24]

In Ettinger's work, psychoanalysis is less intrinsically segregated into registers and stages, and more conceived in terms of a metamorphic relationality.

For Catherine Keller, theology and ontology are essentially relational. Keller is one of our most powerful and creative contemporary theologians, and her work is challenging, provocative, and inspiring. In *Face of the Deep*, Keller rereads the opening chapter of Genesis 1 with the help of biblical scholars; poststructualist philosophers such as Levinas, Derrida, and Deleuze; and Whiteheadian process theologians, to generate a profound theology of becoming. Creation is not *ex nihilo*, as the masculinist theological fantasy would have it, but *ex profundis*, out of a profound and virtually infinite deep. Keller argues that for her tehomic theology, the Derridean "deconstruction of the absolute Logos of the *ex nihilo* yields *an otherness of cosmos bottomlessly preceding and exceeding human language.*"[25]

The Deep is a kind of matrixial borderspace, as is her image of the cloud in *Cloud of the Impossible*. *Cloud of the Impossible* is an extraordinary engagement with negative theology and its "cloud of unknowing," as well as quantum physics with its cloud of entanglement. She uses the theories of Whitehead, Deleuze, and Judith Butler to draw out some of the implications of thinking about reality in terms of folds and foldings, and she reflects on our contemporary situation of precarious life on an unstable planet. Keller's profound apophaticism is also a kind of mourning, a hospice for a potentially dead God and a possibly dying world. Keller says that "the pressure of looming climate cataclysm has been deepening the dangerous denials, the willful ignorance," but she doesn't want apocalypse to have the last word.[26]

How do we think about the end, even if we don't want it to be in apocalyptic terms? Or, as Derrida says, "who, we?" At the end of his essay "The Ends of Man," Derrida argues that we have but two strategies if we want to exit the system of metaphysical humanism that is now experiencing a "total trembling." This text from 1968 evokes both the assassination of Martin Luther King and the insurrection of May '68 in France. Derrida says the first choice is "to attempt an exit and a deconstruction without changing terrain," by resorting to whatever tools are at hand inside, where "we are."[27] The other choice is "to decide to change terrain, in a discontinuous and irruptive fashion . . . by affirming an absolute break and difference." So of course, confronted with an either/or, Derrida refuses to choose. Or rather, he chooses both: "the choice between these two forms of deconstruction cannot be simple and unique." Rather, "a new writing must weave and interlace these two motifs of deconstruction." This new writing is a kind of double writing, which involves a change of style—"and if there is style, Nietzsche reminds us, it must be *plural.*"[28]

I confess I can think of no better example of this new writing for which Derrida calls than Keller's writing and Keller's style in *Face of the Deep* and *Cloud of the Impossible*. This style consists of the interlacing of disciplinary, interdisciplinary, and multidisciplinary texts, including voices within and without "us": our tradition, religion, nation, gender, species, and assemblage. This style of writing that Keller performs for us is a kind of freedom because it subtracts the "we" that exists between the two ends, the first end of the broken source, and the second end of the end of everything, including us, and keeps it in motion, spinning like an entangled word, or particle, or person. It radiates, deconstructively, drawing us up into her cloud like a vortex.

Derrida juxtaposes Nietzsche to Heidegger at the end of "The Ends of Man," and he affirms "the difference between the superior man and the superman." He recalls the end of *Thus Spoke Zarathustra*, where the superior man is "abandoned to his distress in a last movement of pity," while the latter, Zarathustra himself, "awakens and leaves, without turning back to what he leaves behind him. He burns his text and erases the traces of his steps."[29] The laughter of the superman rings out beyond the metaphysical repetition of humanism.

Here, in the "early" Derrida of 1968, the laughter and the affirmation is less edged with nostalgia, loss, and mourning. Later, Derrida will embrace mourning more emphatically. But can we, in fact, choose between celebration and mourning? Keller eloquently offers us both. On the one hand, "gender—and now sex—are so tangled in our queerly eligible Earth that in resonance with an interreligious planetarity their vibrant movements may do much to stir up a sustainable future."[30] On the other hand, "our economic arrangements may undermine that hope," as she quotes Adrian Parr: "We are poised between needing to radically transform how we live and becoming extinct." The middle pose, which is not a middle ground but remains tethered to a relation that yet continues, consists of acknowledging "that radical transformation remains in this time of transition—to put a counterapocalyptic spin on it—maddeningly *possible*."[31]

At the end of "The Ends of Man," Derrida evokes Nietzsche's Superman, who burns his text, turns on his heels, and disappears in a cloud of dust. Now Nietzsche seems just a little too extreme, too hyperbolic, too dated for sophisticated Continental philosophy today. And of course, as we all know, he was an incredible misogynist even if not a racist, an elitist even if not an anti-Semite. Still despite all that we know about Nietzsche, what if—just suppose—that Zarathustra was a woman? What difference would it make? Perhaps none at all.

And yet, still I want to go there, to follow Derrida back to the end of *Thus Spoke Zarathustra* and the cloud of the eternal recurrence. Here Zarathustra has found her long sought-after higher men, and she sings to them the song of how joy wants deep, wants deep eternity. But then the higher men get drunk and go to sleep, and Zarathustra realizes that "they do not understand the signs of my morning; my stride is for them no summons to awaken."[32] So she abandons these higher men to their sleep, and returns to her animals: "You are the right animals for me; I love you. But I still lack the right men."

Thus spoke Zarathustra. And then it happened. She "suddenly heard [her]self surrounded by innumerable swarming and fluttering birds: but the whirring of so many wings and the thronging about [her] head were so great that [she] closed [her] eyes. And verily, like a *cloud* (!) it came over [her], like a cloud of arrows that empties itself over a new enemy. But behold, here it was a cloud of love, and over a new friend."[33] This new friend is a sign that takes the form of a lion. And Zarathustra speaks but a single sentence: "*My children are near, my children.* Then [she] became entirely silent. But [her] heart was loosed, and tears dropped from [her] eyes and fell on [her] hands. And [she] no longer heeded anything and sat there motionless, without warding off the animals any more."[34]

The animals are all over Zarathustra, and she *bears with the chaos*, as Keller would say in *Face of the Deep*.[35] Nietzsche says then that the higher men come out and hear the roar of the lion and then they flee. Zarathustra reflects on what has just happened, and declares that her last temptation was pity for these higher men. She cries out: "My suffering and my pity for suffering—what does it matter? Am I concerned with *happiness*? I am concerned with my *work*."[36] Then she greets the morning, the daybreak. In *Cloud of the Impossible*, Keller invokes the cloud, which is also the cloud of love that appears to Zarathustra, and with it comes the temptation to pity the higher men, the men and women—we—who created this world and this warrior complex and this suffering and death, at least the ones who genuinely understand it and are horrified by it and wish and want to fix it. The higher men are contrasted with the last men, who yawn, don't have a clue to what is happening, and are caught up in climate change denial and voting to elect people chosen by our corporate masters who will accelerate our path to destruction. Zarathustra, however, calls us to a new becoming, another species, and a different end to man, if there is such a thing.

In some ways I am imagining suggesting a genealogy in which Keller becomes a kind of mother figure, taking the place of Nietzsche/Zarathustra, although we have to take great care with such metaphors. She might well not want to adopt me! In addition, while Keller writes very affirmatively about Derrida, she is much more ambivalent about Nietzsche, and she does not mention the cloud that appears at the end of *Thus Spoke Zarathustra*. The point is not so much to affirm this or that philosopher, but to find ways to help us undo the necessary link between fathers and sons in theory and in philosophy.

Theory can never be completely divorced from practice and from life. In an essay included in a book coedited by Keller and Laurel C. Schneider called *Polydoxy*, Colleen Hartung reflects on "Faith and Polydoxy in the

Whirlwind." At the opening of her essay, Hartung recalls a powerful tornado that raged across Nebraska in 1975 while she was in her first year in college. She explains that while she was unhurt, the city of Omaha was cut off and there was no way to communicate with her parents. She relates and reflects upon her mother's statement that her family took the absence of specific news as an encouraging sign that she was not dead or seriously injured. Hartung says that this taking of no news, which could have been anguishing in its uncertainty, for good news, was "an extraordinary act of faith considering that it would have been five years that June since my brother had died of leukemia."[37] This faith on the part of her mother, that "no news was good news," expresses the polydoxy she articulates in theoretical terms, using Derrida and Caputo.

Faith can take place in the absence of God, without God. Hartung evokes the *sans* or "without" that pervades Derrida's work, partly as expressed by John D. Caputo. She says that "Derrida's pursuit of the *sans* provides language that makes a faith without God, that is open to what is wholly other, theoretically unintelligible."[38] It seems to me that religious language *requires* God to be absent, or at least distant, in order to make a faith. Otherwise it wouldn't be faith. That does not mean that God does not or cannot exist, but God cannot collapse the spacing required for faith, and some faiths work with God while other faiths may authentically take place without God. Or at least, we could choose to value these faiths without God as authentic, rather than inauthentic, as lacking, or as being simply and neatly non-religious.

Hartung raises the stakes even further in her essay when she admits that she raids three contemporary theologians for their distinct insights that she then dislocates from the rest of their theological edifices, and then assimilates them to her own theological vision, which is a theology without God. What is more, she chooses to elaborate on these three themes of blindness, embrace, and courage by way of an appropriation of and separation from three of the most creative and important theologians working and writing today, and not only that, from three theologians whose chapters are also included in the book *Polydoxy*, including the coeditors of the book. These two coeditors are Laurel Schneider and Catherine Keller, and the third theologian is Mayra Rivera, author of *The Touch of Transcendence* and *Poetics of the Flesh*.

Schneider is a theologian of multiplicity, and dislodging the logic of the One allows her to see divinity in its manifestation as multiplicity. Hartung claims that for Schneider, "to face and to see the fragile is to realize divine occurrence as incarnation again and again."[39] By contrast, Hartung remains faithful to the blindness inherent in her mother's faith and in

Derrida's thought that paradoxically sustains faith. According to Hartung, "Schneider's emphasis on presence and on seeing . . . considered from a Derridean perspective, might collapse this space between believing and seeing that Derrida attempts to keep open as a way of holding at bay a totalizing foreclosure that a too optical metaphysics of presence effects."[40] If Hartung is correct, then Schneider twists free of the logic and metaphysics of the One, but perhaps her theology is still too entranced with a metaphorics and metaphysics of vision. Despite this distancing from Schneider's affinity for seeing, Hartung celebrates Schneider's foregrounding of ambiguity in bodies and in faith.

From Rivera, Hartung takes the phenomenality of touch or even embrace. Faith is a risky embrace of what is uncertain and unknown. But Rivera chooses to hold onto a transcendence in which to anchor her understanding of faith as touch. For Hartung, she herself chooses instead to remain faithful to the faith of her parents after the death of her brother, who "let go of the name of God and yet in that release realized a faith that could and would, again and again, turn toward an unthinkable, unspeakable, ungraspable possibility—the death of a child."[41] The death of one's child would seem to be one of the most faith-shattering events that could ever occur.

According to Hartung, Rivera wants to limit Derridean deconstruction by tracing its originary source to divine transcendence on the other side of *différance*. Hartung, however, wants to remain with *différance* and its indeterminacy all the way. This is a cryptic move, just as the saying "no news is good news" is a cryptic saying, but "perhaps it could be read as a disclosure, as an opening that might make a space for love."[42] Do we need God or divinity to make love with and as bodies, or does the heaviness of God's transcendence weigh love down? Hartung uses the image of a surrender of arms; she evokes the notion of giving up our arms and embracing life without the conceptual tools we generally employ to keep it in check. "Here the surrender of arms is an embrace," she says, "without the benefit of technological or religious good news."[43] Technological and religious arms are deployed to overdetermine life, to make it make sense, to make it appear good, to instrumentalize our experience and protect us from the uncertainty that makes living worthwhile. But these arms ward off the possibility of the new.

With Keller, finally, Hartung is more tentative than with Rivera or Schneider. Hartung appreciates and appropriates Keller's courage in the face of the deep, which is also the tehomic fury of the whirlwind. The only thing she lets go of that Keller does not is the issue with which Hartung begins this section of encounters, which is the name of God. Hartung confesses: "for me, the tehomic rhythm of life has washed away any name that would underwrite faith with a destinal assurance. Yet this has not left me faithless."[44] Here there is a tension because while Keller does not give up the name of God, she also does not provide it any destinal assurance. Furthermore, at least Caputo, if not Derrida, holds onto the name of God, but also in a way that would not guarantee or underwrite faith in any substantial way. Derrida is more complicated because in his work he both saves and gives up the name of God, and the *sans* is not a simple without, but always both a with and a without. This is a paradox, but then so is faith.

The real issue here concerns Hartung's provocative claim, which she expresses in relation to both Schneider and Keller (and Rivera more implicitly): "In its polydoxy, theology risks a consideration of faith without the name of God."[45] Hartung quotes Keller as raising the relevant question: "What would theology be *without* theos?" (161). Keller does not answer this question, according to Hartung, but Hartung does suggest, with Derrida, that alternative theological beginnings without God could resonate "with a faithful Derridean deconstructive posture that courageously turns toward the unforeseeable."[46] The passion of a faith without God no longer knows to whom or what to pray. One of the effects of moving between determinate faiths is a wearing away of any pure, authentic, or determinate lineage, and Hartung evokes this experience both for herself and for Derrida. She says that Derrida's "is not a legacy of faith inscribed in a determinable lineage that might be Christian or Jewish or democratic. Rather this inheritance is a plea . . ." (162). So the tehomic rhythms that ebb and flow may wear down the sharp edges of determinate lineages, and the lines may blur such that we do not know not only to whom we pray but who we are when we pray or plead. And this is a complex inheritance, which is incarnated in Hartung's mother and her particular faith.

For Hartung, her mother's theology, encapsulated in the phrase "no news is good news," "resists gestures of faith that make light of a darkness, a blindness, and a grief that is the undoing of oneself and of the world over and over again."[47] We do not often get good news, but sometimes no news is the best news we could ever receive, and most of the time we have no idea how fortunate we are not to get news of disaster, illness, or the death of a loved one. What Hartung performs in this extraordinary essay is an existential theology that is not conceptually impoverished, but one that finds a way to honor both Derrida and her mother by threading her theology through the eye of the needle of three polydox theologians and stitching together something both refreshing and common. This faith of Hartung's mother as elaborated through Derrida "insinuates rather than claims. It holds on for dear life," caught up amidst the whirlwind.[48]

My mother is not a philosopher. Both of my parents are incredibly smart; however, my mother was not given the same level of encouragement and esteem as my dad got. But there is a fundamental difference in how I can relate to them intellectually. My father is supportive and proud of my success, and we now have a very positive relationship, but he does not really understand my philosophical and theoretical bent. He is much more pragmatic in his outlook. My mom, on the other hand, has always profoundly appreciated and resonated with my ideas, has always been able to understand me and to encourage me even when she didn't know the philosophical background about the theories that I was talking about. That is, she could and can grasp the most complex and profound ideas, intuitively, despite not having studied them. This connection, which also exists between me and her brother, my uncle, has given me so much inspiration and confidence in my pursuit of my intellectual vocation. It also has helped me understand how people can have the capacity to think philosophically even without any conventional study or explicit training, as a more democratic practice as opposed to a more stereotypical elitism.

One of the ways I try to relate to both of my parents is to be self-sufficient enough so that they get no bad news, and this lack of news is good news, as with Hartung and her mother. I love my parents in a way that wants to not cause them worry and anxiety, although I know that any caring relationship necessarily entails such worry. One of the elements of generation, like every relation to another living being, involves a relation to death and to mourning. When you have a child, one of your hopes is that you do not outlive your children. When you are a child, one of the realizations of growing up is accepting that your parents do not want to outlive you, even if they want to live as long as possible.

When someone close to you dies, you grieve and suffer and mourn. When someone you don't know dies, you don't usually feel any direct experience of loss. Love involves the possibility of mourning the loss of that love and that person, literally or not. Mourning is an important theme of Derrida's, as many readers have observed. In his book *Not Half No End*, Geoffrey Bennington says that one thing Derrida taught him is that "life is an economy of death."[49] Bennington explains that for Derrida, mourning is an incomplete, interrupted process that he sometimes called "half-mourning," or *demi-deuil*.[50] This demi-deuil, or incomplete mourning, is connected with melancholia, but not simply as a pathological condition. For Derrida, mourning and melancholia structure our experience of life, not as negative circumstances that we get over, but as ongoing, endless processes.

So long as we live, we mourn. Or half-mourn to no end. In a powerful reflection on loss and love, Katerina Kolozova mourns the death of her father. Although her discussion draws mostly on François Laruelle, I think that her writing also has resonances with Derrida's work. In her chapter "The Real Transcending Itself (Through Love)," which is included in her book *Cut of the Real*, Kolozova explains that she takes a heretical position vis-à-vis poststructuralist feminism. For her, the non-philosophy developed by Laruelle means that she is free from any philosophical orthodoxy, or obligation to treat philosophical materials in an already structured, systematic manner. All philosophical concepts form a realm of *chora*, a "domain of transcendental material" from which to draw and make use.[51] For non-philosophy, philosophical language cannot capture the real; all that we can do is posit the real in a "vision-in-one," and then work backwards from there to our own experience.

For Kolozova, language is inadequate to the real, but it can correlate with the resonance of the real in our lives. The concept of the real is used by Laruelle, but it is also fundamentally inflected by Lacan's theorization of it as something inaccessible and inexpressible in symbolic terms. We cannot fully express the real in language, but in specific situations language can resonate with the real. When this occurs, "satisfaction is taken in the form of the token of this 'love': the translucent texture of language, weaving around the thickness of experience."[52] We experience ourselves as uniquely human, or what Laruelle calls "the identity in the last instance of the human-in-human," as a state of radical solitude. According to Kolozova, love offers a way to think about the relation between language and the real, which is an asymmetrical relationship. We do not think from our language to the real; we can only go from the positing of the Real back to our lived experience of language. Laruelle attempts to formalize an insight into our experience of the Real in a way that works against the presumed mastery and sufficiency of philosophy. Kolozova adopts Laruelle's approach and also radicalizes it with her insights into feminism.

It is impossible to speak of love. It is impossible to go directly from our solipsistic self to an other. So much of what is thought under the name of poststructuralism involves underlining and complicating this relationship between self and other. Love is impossible, but for Kolozova "an act of love" involves "an act of attempting to reach out to the other as the instance of salvation from one's radical self-enclosure."[53] Derrida would say that we are always already inhabited by an Other, whereas Laruelle and Kolozova use a different language. Kolozova claims that we are inhabited by a Stranger, which correlates with the real.

There is a fluidity of the body in the philosophy of Luce Irigaray and a materialism of becoming in the philosophy of Rosi Braidotti that Kolozova both affirms and radicalizes. Becoming takes place in the flesh, in the lived experience of our material existence. At the same time, the effect of these fluid, material and embodied processes produces the radical solitude that defines our identity as human in the last instance. As Kolozova states, "we were all 'born'—as an 'I,' as that most primitive sense of self-hood—in the most immediate experience of the state of inescapable situatedness in ourselves."[54] This means that the possibility of our ability to reflect upon ourselves is tied to the very impossibility of reflection, of the existence that escapes reflection, which is an acutely Derridean theme.

According to Kolozova, the mediation of the real "is enabled by and originates from the experience of radical solitude." The mediation of the real by the experience of radical solitude paradoxically produces the experience of love, "which finds itself at the very origin of all and any transcendence, at the heart of the creation of all our world(s)."[55] Here language is not simply impotent because it effects this experience of love, which is also an expression of empathy. Kolozova forges a new language of love, that works with and through the experience of radical solitude that is quintessentially our own, to find a way to share with the other. As she explains:

> Empathy with the radical solitude of the other by way of identifying one's own enacted state of radical solitude with that of the (imagined) other is an *act* made both of the event of "having lived through" (*le vécu*) and of the event of the mediation provided by language (or the transcendental).[56]

We connect our own experience of radical solitude with that of the other, and thus generate empathy, compassion, or love.

The Stranger in us bridges the gap between our radical and unbearable solitude and that of the other which is her Stranger. This connection is a fashioning of an unstable solidarity, one that acknowledges the vulnerability of thought that faces "its own desire for (or the fantasy of) the ultimate truth as it is for the impossible immediacy of the other's real."[57] The death of the other brings us back to that originary solitude. The experience of mourning the death of the beloved involves "the presentification of the real in its absolute form, cleansed from the soothing, that is, *estranging*, workings of the world."[58] Here mourning is an experience that connects us to the real in a very powerful and particular way.

We cannot get to the real from the standpoint of language. We cannot adequately describe the real using philosophical language. We can, however, go from the real, posited as a "vision-in-one," to language, and through language to our lives and our lived experience. The mourning of the loved one involves the survival of a phantasm of this person in our sensations. The phantasm is, in many ways, imaginary, but it does express the real.

Kolozova says that our remembrance of the lost person is "projected by the dark, thick sensation of the real of love and the real of loss. The event of longing for this . . . loved one (*in-dividual*) is real; it is in the real."[59] Our remembrance, which is filtered through sensations of pain and nostalgia expresses the real in the world in way that recreates and renews love. As Kolozova concludes, "the real (of the sheer trauma) and the thought (of the world of language) are the coagulating elements of this emerging brilliance of renewed desire for the other—of the love born anew."[60]

I think that Kolozova's analysis of love and of mourning is rich and insightful; I also think that is resonates with my experience of Derrida. Specifically, the fact that she is mourning her father is significant because I am wrestling with my love for Derrida as a person and as a philosopher, as well as with my ambivalence concerning this sexist situation because Derrida also is a father-figure. Perhaps it is healthier to mourn the death of the father in feminist terms than it is to try to save or redeem the father. Or to try to put the father to death oneself, which may be too oedipal, and simply risks repeating the masculinist re-appropriation of the father. I don't have the answer; as a white male I am also part of the problem. But the rebellious narcissism of Vahanian, the transgressing of gender at the heart of being of Malabou, and the radical solitude and love of Kolozova are profoundly encouraging to me, as are the cloud and the deep of Keller. Furthermore, I think that each draws out something that is more implicit in Derrida, that Derrida does not and cannot express.

The love and mourning of fathers and mothers pushes us to and hopefully in some respects beyond the limits of filiality, and this reflection also reveals another scene where Derrida's work is relevant beyond any simple conception of writing. In this book, I have tried to think with and about Derrida in terms that are not simply faithful, but in ways that demonstrate how his philosophy still matters in a number of ways, for religion, politics, materialism, ethics, science, and life. Love is always a form of narcissism, as well as a form of betrayal. Absolute fidelity is impossible, but we can acknowledge and embrace these efforts to understand and work through our thinking and others', which includes others' thinking as our own. Derrida continues to inhabit and possesses me in a powerful way, and here I have tried to give voice to this asymmetrical relationship.

Perhaps someday in the wake of metaphysics, at the end of patriarchy, there will be a plastic dawn. And in that dawn there will be a new kind of book, a profoundly Malabouian book, assuming anyone remains capable of reading books. Such a book might be called *Energy and Change*. And it would be a fundamentally Derridean book to come.

Acknowledgments

An earlier version of Chapter 2 was published as "Surviving Christianity" in *Derrida Today* 6.1 (2013), 23–35, and it has been revised for this book. Thanks to Edinburgh University Press for permission to republish it here.

An earlier version of Chapter 3 was published as "Political Theology Without Sovereignty: Reading Derrida Reading Religion," in *Politics and Religion*, ed. Saitya Brata Das (Delhi, Aakar Books, 2014). It has been substantially revised and expanded for this book.

An earlier version of Chapter 4 was published as "Interrupting Heidegger: Celan's Poetry in Derrida's Thought," in *Paul Celan: Da Ética do Silêncio à Poética do Encontro*, ed. Cristina Beckert, Maria João Cantinho, Carlos João Correia, and Ricardo Gil Soeiro (Lisboa: Centro de Filosofia da Universidade de Lisboa, 2014). It has been substantially revised and expanded for this book.

I want to thank Tom Lay and everyone else at Fordham University Press who supported and encouraged me on this project, including two readers for the Press who contributed excellent critical and constructive feedback. In particular I appreciate the incredible work and vision of the late Helen Tartar, who told me that she wanted to publish this book when I first began imagining it in 2012. Her memory and her influence continue for me and for everyone she touched and with whom she worked.

I would also like to thank everyone who has helped me read and think about Derrida over the past twenty-five years, starting with a course I took at William and Mary in "Special Topics in Linguistics: Language, Meaning and Theory from the Bible to Derrida," with Talbot Taylor and Colleen Kennedy, and continuing with a course I took during my M.A. program in Religious Studies at the University of Virginia on hermeneutics, taught by Larry Bouchard. In both courses, we read parts of *Of Grammatology*. Other influences include the late Charlie Winquist, Catherine Keller, Mark C. Taylor, Gregg Lambert, Catherine Malabou, Carl Raschke, Karen Barad, Richard Kearney, Yvonne Sherwood, Stephen Shakespeare, and Kevin Hart.

I have also benefited from discussions and conversations with and feedback from Jeff Robbins, Noëlle Vahanian, Creston Davis, Karen Bray, Ward Blanton, Taine Duncan, Satya Das, Franson Manjali, Mary-Jane Rubenstein, Mike Grimshaw, Santiago Zabala, Kevin Mequet, Malik Saafir, Guo-ou Zhuang, Zhaoguo Dai, Charlie Harvey, Ricardo Gil Soeiro, Maria João Cantinho, Philip Goodchild, Jim DiCenso, and Victor Taylor.

Students in my classes have helped me immensely, even as they often struggled along with me to understand Derrida's writings; thanks go to Tim Snediker, Keith Witty, Sara Harvey, Mason Brothers, Savannah Moix, Jessica Thames, Keith Dove, Matthew Spencer, Chloe Zedlitz, Emily Jones, Ashley Hunter, Jonathan Bowman, Aaron Mark, Holly Hughes, Maeghan McClure, Simon Stone, Janie Brown, Payton Collier, Amanda Clark, Richie Kearney, and Payton Peebles-Hartford. My apologies to anyone I left out.

Most of all, I want to thank Jack Caputo, for his help, his friendship, his pioneering work on Derrida, and his own ongoing vital engagements with Continental philosophy and radical theology.

Notes

Introduction: Derrida and the New Materialism

1. See Michael Naas, *Miracle and Machine: Jacques Derrida and the Two Sources of Religion, Science, and the Media* (New York: Fordham University Press, 2012). Naas interprets Derrida's entire philosophy through an extensive reading of Derrida's most explicit engagement with religion, the essay "Faith and Knowledge: The Two Sources of Religion at the Limits of Reason Alone." The "miracle" applies to a form of exceptional and singular ethical responsibility that animates life and supplies it with dignity, while the "machine" indicates the inescapable technical repetition of life that exposes and delivers it to death. There is no miracle without a machine, and yet the machine cannot foreclose the "miracle" of responsibility, of being capable of responding to an other. Derrida suggests in much of his later work that not only humans are capable of responding; animals can as well.

2. See Arthur Bradley, *Originary Technicity: The Theory of Technology from Marx to Derrida* (New York: Palgrave Macmillan, 2011).

3. In addition to specific works by these authors, see the volumes *New Materialisms: Ontology, Agency, Politics*, ed. Diana Coole and Samantha Frost (Durham: Duke University Press, 2010); and Rick Dolphijn and Iris van der Tuin, *New Materialism: Interviews & Cartographies* (Michigan: Open Humanities Press, 2012). A great resource on Malabou, which treats her work from a materialist perspective, is *Plastic Materialities: Politics, Legality, and Metamorphosis in the Work of Catherine Malabou*, ed. Brenna Bhandar and Jonathan Goldberg-Hiller (Durham: Duke University Press, 2015).

4. Jacques Derrida, *Writing and Difference*, trans. Alan Bass (Chicago: University of Chicago Press, 1978), 151.

5. See Clayton Crockett and Jeffrey W. Robbins, *Religion, Politics, and the Earth: The New Materialism* (New York: Palgrave Macmillan, 2012).

6. "Interview with Rosi Braidotti," in Dolphijn and van der Tuin, *New Materialism*, 21.

7. Vicki Kirby, *Quantum Anthropologies: Life at Large* (Durham: Duke University Press, 2011), 8.

8. Ibid., 133.

9. Michael Barnes Norton, "Matter and Machine in Derrida's Account of Religion," *Sophia*, DOI 10.1007/s11841-014-0452-y, December 2014.

10. John D. Caputo, *The Insistence of God: A Theology of Perhaps* (Bloomington: Indiana University Press, 2013), 170.

11. Karen Bray, "Becoming Feces: New Materialism and the Deep Solidarity in Feeling Like Shit," in *Religious Experience and New Materialism: Movement Matters*, ed. Joerg Rieger and Edward Waggoner (New York: Palgrave Macmillan, 2015), 128–29.

12. Ilya Prigogine and Isabelle Stengers, *Order Out of Chaos: Man's New Dialogue With Nature* (New York: Bantam Books, 1984), 143.

13. Derrida, *Writing and Difference*, 25.

14. Ibid., 27.

15. Carl A. Raschke, *Force of God: Political Theology and the Crisis of Liberal Democracy* (New York: Columbia University Press, 2015), 36.

1. Reading Derrida Reading Religion

1. See Benoît Peeters, *Derrida: A Biography*, trans. Andrew Brown (Cambridge: Polity Press, 2013), 503.

2. For other readings of Derrida's engagements with religion, see Michael Naas, *Miracle and Machine: Jacques Derrida and the Two Sources of Religion, Science, and the Media* (New York: Fordham University Press, 2012); Steven Shakespeare, *Derrida and Theology* (London: Continuum, 2009); Kevin Hart, *The Trespass of the Sign: Deconstruction, Theology and Philosophy* (Cambridge: Cambridge University Press, 1989); *Derrida and Religion: Other Testaments*, ed. Yvonne Sherwood and Kevin Hart (London: Routledge, 2005); and Hent de Vries, *Philosophy and the Turn to Religion* (Baltimore: Johns Hopkins University Press, 1999).

3. Edward Baring, *The Young Derrida and French Philosophy, 1945–1968* (Cambridge: Cambridge University Press, 2011), 5.

4. On the growing awareness of the significance of Althusser's early Catholicism on his later philosophy, see Roland Boer, "Althusser's Catholic Marxism," *Rethinking Marxism: A Journal of Economics, Culture, and Society*, Vol. 19, No. 4, 2007, 469–86.

5. See ibid., Chapter 3, 82–112.

6. On Derrida's relation to Judaism, see *Judeities: Questions for Jacques Derrida*, trans. Bettina Bergo and Michael B. Smith (New York: Fordham University Press, 2007).

7. See Gayatri Chakravorty Spivak, *A Critique of Postcolonial Reason: Toward a History of Our Vanishing Present* (Cambridge: Harvard University Press, 1999), 425.

8. Steven Shakespeare, *Derrida and Theology*, 197.

9. Carl A. Raschke, *Force of God: Political Theology and the Crisis of Liberal Democracy* (New York: Columbia University Press, 2015), 16–17.

10. Jacques Derrida, *Edmund Husserl's Origin of Geometry: An Introduction*, trans. John P. Leavey, Jr. (Lincoln: University of Nebraska Press, 1989), 148 (emphases in original).

11. Jacques Derrida, *Of Grammatology*, trans. Gayatri Chakravorty Spivak (Baltimore: Johns Hopkins University Press, 1976), 11.

12. Ibid., 22–23 (emphasis in original).

13. Ibid., 23.

14. Edward Baring points out that in his early work on Husserl, including his introduction to Husserl's *Origin of Geometry*, Derrida uses the term "difference" but has not yet fully distinguished it from Heidegger's ontico-ontological difference. See Baring, *The Young Derrida*, 191.

15. Jacques Derrida, *Margins of Philosophy*, trans. Alan Bass (Chicago: University of Chicago Press, 1982), 8.

16. Jacques Derrida, *Writing and Difference*, trans. Alan Bass (Chicago: University of Chicago Press, 1978), 278.

17. Ibid., 289.

18. See John D. Caputo, *The Insistence of God: A Theology of Perhaps* (Bloomington: Indiana University Press, 2013).

19. Derrida, *Writing and Difference*, 83.

20. Ibid.

21. Ibid., 96 (emphasis in original).

22. Ibid., 116.

23. Ibid., 131.

24. Jacques Derrida, "How to Avoid Speaking: Denials," in *Derrida and Negative Theology*, ed. Harold Coward and Toby Foshay (Albany, SUNY Press, 1992), 103.

25. See ibid., 110.

26. More recently, we can see a recurrence of the Nazism that attends to any consideration of the significance of Heidegger's thought in the publication of the "Black Notebooks" from 1931 to 1941, with their sometimes direct expressions of antisemitism. See Martin Heidegger, *Ponderings II–VI: Black Notebooks 1931–38*, trans. Richard Rojcewicz (Bloomington: Indiana University Press, 2016).

27. Jacques Derrida, "Force of Law: The Mystical Foundation of Authority," in *Acts of Religion*, ed. Gil Anidjar (New York: Routledge, 2002), 243 (emphasis in original).

28. Ibid. (emphasis in original).

29. Walter Benjamin, "Critique of Violence," in *Selected Writings, Volume I, 1913–1926*, ed. Marcus Bullock and Michael W. Jennings (Cambridge: Harvard University Press, 1996), 252.

30. Jacques Derrida, *The Gift of Death*, trans. David Wills (Chicago: University of Chicago Press, 1995), 68.

31. Ibid., 82.

32. Ibid., 84.

33. Ibid.

34. Jacques Derrida, *Specters of Marx: The State of the Debt, the Work of Mourning, and the New International*, trans. Peggy Kamuf (London: Routledge, 2004), 65.

35. Ibid.

36. Jacques Derrida, "Faith and Knowledge: The Two Sources of Religion at the Limits of Reason Alone," trans. Samuel Weber, in *Religion*, ed. Jacques Derrida and Gianni Vattimo (Stanford: Stanford University Press, 1998), 33 (emphasis in original).

37. Ibid., 44.

38. Ibid., 51.

39. Ibid.

40. Ibid., 63.

41. Naas, *Miracle and Machine*, 94.

42. Ibid., 95.

2. Surviving Christianity

1. Jacques Derrida, *The Post-Card: From Socrates to Freud and Beyond*, trans. Alan Bass (Chicago: University of Chicago Press, 1987), 394 (emphasis in original).

2. Gayatri Chakravorty Spivak, *A Critique of Postcolonial Reason: Toward a History of the Vanishing Present* (Cambridge: Harvard University Press, 1999), 425.

3. Jacques Derrida, "Faith and Knowledge: The Two Sources of 'Religion' at the Limits of Reason Alone," in *Religion*, ed. Jacques Derrida and Gianni Vattimo (Stanford: Stanford University Press, 1998), 5. See also Michael Naas's discussion of this important text in *Miracle and Machine: Jacques Derrida and the Two Sources of Religion, Science, and the Media* (New York: Fordham University Press, 2012).

4. See Jacques Derrida, *Spurs: Nietzsche's Styles*, trans. Barbara Harlow (Chicago: University of Chicago Press, 1981).

5. Derrida, "Faith and Knowledge, 5, 33 (emphasis in original).

6. Ibid., 5.

7. Jean-Luc Nancy, *Dis-Enclosure: The Deconstruction of Christianity*, trans. Bettina Bergo, Gabriel Malenfant, and Michael B. Smith (New York: Fordham University Press, 2008), 142.

8. Ibid., 143.

9. Ibid.

10. Ibid.

11. Ibid., 148.

12. Ibid., 149

13. See Réné Girard, *Violence and the Sacred*, trans. Patrick Gregory (Baltimore: Johns Hopkins University Press, 1979), and *Things Hidden Since the*

Foundation of the World, trans. Stephen Bann and Michael Metteer (London: Continuum, 2003).

14. Ibid., 140 (emphasis in original).

15. Ibid.

16. Jacques Derrida, *On Touching—Jean-Luc Nancy*, trans. Christine Irizarry (Stanford: Stanford University Press, 2005), 54.

17. Ibid., 60.

18. Ibid.

19. Ibid., 220.

20. Ibid., 257.

21. Ibid., 257–58.

22. Ibid., 261 (emphasis in original).

23. Ibid.

24. Ibid., 262.

25. Jacques Lacan, *The Triumph of Religion, Preceded by Discourse to Catholics*, trans. Bruce Fink (Cambridge, UK: Polity Press, 2013), 63.

26. Walter Benjamin, "Capitalism as Religion," in *Walter Benjamin: Selected Writings, Volume 1, 1913–1926*, ed. Marcus Bullock and Michael W. Jennings (Cambridge: Harvard University Press, 1996), 290.

27. Naas, *Miracle and Machine*, 95.

28. John D. Caputo, *The Prayers and Tears of Jacques Derrida: Religion Without Religion* (Bloomington: Indiana University Press, 1997), 62.

29. Ibid., 222.

30. Ibid., 289.

31. See John D. Caputo, *The Weakness of God: A Theology of the Event* (Bloomington: Indiana University Press, 2006), as well as *The Insistence of God: A Theology of Perhaps* (Bloomington: Indiana University Press, 2013).

32. Jacques Derrida, "Above All No Journalists!" trans. Samuel Weber, in *Religion and Media*, ed. Hent de Vries and Samuel Weber (Stanford: Stanford University Press, 2001), 69. Quoted in Naas, *Miracle and Machine*, 95–96.

33. Martin Hägglund, *Radical Atheism: Derrida and the Time of Life* (Stanford: Stanford University Press, 2008), 121.

34. Ibid., 34.

35. Ibid., 121 (emphasis in original).

36. Ibid.

37. Ibid.

38. Ibid., 127.

39. Ibid.

40. See the exchange between Caputo ("The Return of Anti-Religion: From Radical Atheism to Radical Theology") and Hägglund ("The Radical Evil of Deconstruction: A Reply to John Caputo") in the *Journal for Cultural and Religious Theory*, Vol. 11. No. 2, 2011, http://www.jcrt.org/archives/11.2/index .shtml. See also Daniel M. Finer's critique of Hägglund's valorization of

survival as a form of narcissism in the same issue: "Radical Narcissism and the Freedom to Choose Otherwise: A Critique of Hägglund's Derrida."

41. Jacques Derrida, "What Is a 'Relevant' Translation?" trans. Lawrence Venuti, *Critical Inquiry* 27, no. 2, Winter 2001), 174–200, quote 174.

42. Ibid., 184.

43. Ibid.

44. Ibid., 186.

45. Ibid., 191.

46. Ibid., 194.

47. Ibid., 197.

48. Ibid. (emphasis in original).

49. Ibid.

50. Ibid., 198.

51. Ibid., 199.

52. Ibid.

53. Ibid.

54. Ibid.

55. Gil Anidjar, *Blood: A Critique of Christianity* (New York: Columbia University Press, 2014), 17.

56. Ibid., 85.

57. Ibid.

58. Ibid., 38.

59. Ibid., 133.

60. Ibid., 134.

61. Ibid., 41.

62. See J. Kameron Carter, *Race: A Theological Account* (Oxford: Oxford University Press, 2008).

63. Anidjar, *Blood*, 49, 45.

64. Ibid., 63.

65. Nelson Maldonado-Torres, "Religion, Conquest, and Race in the Foundations of the Modern/Colonial World," *Journal of the American Academy of Religion* 82, no. 3 (September 2014),636–65, quote 646.

66. Ibid., 151.

67. Ibid., 152.

68. Ibid., 153.

69. Derrida, "Faith and Knowledge," 66 (emphasis in original).

70. Derrida wrote his final words on an envelope for his son to read at his graveside in October 2004, concluding with the sentence: "I love you and am smiling at you from wherever I am." See Naas, *Miracle and Machine*, 269.

3. Political Theology Without Sovereignty

1. Michael Naas, *Miracle and Machines: Jacques Derrida and the Two Sources of Religion, Science, and the Media* (New York: Fordham University Press, 2012), 97.

2. Ibid., 118.

3. See Arthur Bradley, *Originary Technicity: The Theory of Technology from Marx to Derrida* (Hampshire, UK: Palgrave Macmillan, 2011).

4. Jacques Derrida, "Faith and Knowledge: The Two Sources of 'Religion' at the Limits of Reason Alone," in *Religion*, ed. Jacques Derrida and Gianni Vattimo (Stanford: Stanford University Press, 1996), 50.

5. Steven Shakespeare, "The Persistence of the Trace: Interrogating the Gods of Speculative Realism," in *The Future of Continental Philosophy of Religion*, ed. Clayton Crockett, B. Keith Putt, and Jeffrey W. Robbins (Bloomington: Indiana University Press, 2014), 85–86.

6. Ibid., 51.

7. Naas, *Miracle and Machine*, 275.

8. Ibid., 191.

9. For an update of Schmitt, with an application of his theology to the American context with its political imaginary of sacrifice, see Paul W. Kahn, *Political Theology: Four New Chapters on the Concept of Sovereignty* (New York: Columbia University Press, 2011).

10. Roberto Esposito, *Two: The Machine of Political Theology and the Place of Thought*, trans. Zakiya Hanafi (New York: Fordham University Press, 2015), 3.

11. Ibid., 21.

12. Ibid., 20.

13. See Jacques Derrida, *Dissemination*, trans. Barbara Johnson (Chicago: University of Chicago Press, 1983).

14. Derrida, "Faith and Knowledge," 51.

15. Ibid.

16. Geoffrey Bennington, *No Half No End: Militantly Melancholic Essays in Memory of Jacques Derrida* (Edinburgh: Edinburgh University Press, 2010), 31.

17. Carl Schmitt, *Political Theology: Four Chapters on the Concept of Sovereignty*, trans. George Schwab (Chicago: University of Chicago Press, 2005), 36.

18. Ibid., 65.

19. Ibid., 5.

20. Ibid., 65.

21. See Carl Schmitt, *The Concept of the Political*, trans. George Schwab (Chicago: University of Chicago Press, 1996), 26: "The specific political distinction to which political actions and motives can be reduced is that between friend and enemy."

22. Jacques Derrida, *The Politics of Friendship*, trans. George Collins (New York: Verso, 1997), 84, 243.

23. Ibid., 68 (emphasis in original).

24. Ibid.

25. Ibid., 69.

26. Ibid.

27. Saitya Brata Das, *The Wounded World: Essays on Ethics and Politics* (Delhi: Aakar Books, 2013), 14.

28. Jacques Derrida, *Specters of Marx: The State of the Debt, The Work of Mourning, and the New International*, trans. Peggy Kamuf (New York: Routledge, 1994), 167.

29. Jacques Derrida, *The Beast and the Sovereign*, Volume I, trans. Geoffrey Bennington (Chicago: University of Chicago Press, 2009), 75.

30. Ibid., 76–77.

31. Ibid., 282.

32. Schmitt, *Political Theology*, 62.

33. See Alain Badiou, *Metapolitics*, trans. Jason Barker (New York: Verso, 2005), 20.

34. Jacques Derrida, *Rogues: Two Essays on Reason*, trans. Pascale-Anne Brault and Michael Naas (Stanford: Stanford University Press), 13.

35. Ibid., 16.

36. See ibid., 18.

37. Ibid., 82.

38. Ibid., 110.

39. Ibid., 77. See my suggestion that this god could be viewed less in terms of Judeo-Christian divinity than as an African or Caribbean *orisa* or *lwa*, in "Vodou Economics: Haiti and the Future of Democracy," in *Deleuze Beyond Badiou: Ontology, Multiplicity and Event* (New York: Columbia University Press), 185–94.

40. Ibid., 114.

41. In *Rogues*, Derrida distinguishes "a god" from both "the One God" and "gods" in his reflections on Heidegger (110). In *Deleuze Beyond Badiou*, I ask, "What if democracy involves serving the lwa, becoming the horse to assist in serving the people?" (193).

42. Jeffrey W. Robbins, *Radical Democracy and Political Theology* (New York: Columbia University Press, 2011), 113.

43. Antonio Negri, *Spinoza for Our Time: Politics and Postmodernity*, trans. William McCuaig (New York: Columbia University Press, 2013), 32.

44. See Paul Kahn, *Political Theology: Four New Chapters on the Concept of Sovereignty* (New York: Columbia University Press, 2011). Kahn updates Schmitt and applies his work to the United States. Kahn's understanding of political theology is based ultimately on the idea of sacrifice, but here sacrifice reinforces rather than dismantles sovereignty.

45. Jacques Derrida, "Taking a Stand for Algeria," trans. Boris Belay, in *Acts of Religion*, ed. Gil Anidjar (London: Routledge, 2002), 301–8, quote 306.

46. John D. Caputo, *The Weakness of God: A Theology of the Event* (Bloomington: Indiana University Press, 2006), 39.

47. John D. Caputo, *The Insistence of God: A Theology of Perhaps* (Bloomington: Indiana University Press, 2013), 4.

48. Partha Chatterjee, *The Black Hole of Empire: History of a Global Practice of Power* (Princeton: Princeton University Press, 2012), 336. See also Arvind-Pal S. Mandair's decolonization of postsecular philosophy in *Religion and the*

Specter of the West: Sikhism, India, Postcoloniality, and the Politics of Translation (New York: Columbia University Press, 2009). Mandair attends to the postcolonial elements of Derrida's theory, showing how religion works along with colonial power to construct "globalatinization," at the heart of which lies a "structure or belief that Mandair calls "the *global fiduciary*" (420).

49. Chatterjee, *The Black Hole of Empire*, 343–44.

50. Achille Mbembe, *On the Postcolony* (Berkeley: University of California Press, 2001), 78.

51. Ibid., 79.

52. Chatterjee, *The Black Hole of Empire*, 344.

53. See Thomas Picketty, *Capital in the Twenty-First Century*, trans. Arthur Goldhammer (Cambridge: Harvard University Press, 2014).

54. Michael T. Klare, *The Race for What's Left: The Global Scramble for the World's Last Resources* (New York: Metropolitan Books, 2012), 18.

55. Derrida, *The Beast and the Sovereign*, Volume 1, 17.

56. Jacques Derrida, *The Beast and the Sovereign* Volume II, trans. Geoffrey Bennington (Chicago: University of Chicago Press, 2011), 104. See also Michael Naas, *The End of the World and Other Teachable Moments: Derrida's Final Seminar* (New York: Fordham University Press, 2015).

57. Ibid., 105, 255.

58. Ibid., 256.

59. Ibid., 263.

60. Ibid., 264.

61. Ibid., 266.

62. Ibid., 9.

63. Ibid., 279.

64. Ibid., 288.

65. Ibid., 290.

4. Interrupting Heidegger with a Ram: Derrida's Reading of Celan

1. See the discussion by Pierre Joris, "Celan/Heidegger: Translation at the Mountain of Death," 1988, http://wings.buffalo.edu/epc/authors/joris/todtnauberg.html.

2. Paul Celan, "Todtnauberg," in *Poems of Paul Celan*, trans. Michael Hamburger (New York: Persea Books, 2002), 281.

3. Ibid.

4. Ibid.

5. Han-Georg Gadamer, *Philosophical Apprenticeships*, trans. Robert R. Sullivan (Cambridge: MIT Press, 1985), 55.

6. Ibid.

7. We should be careful not to over-emphasize the singularity of this encounter because Celan and Heidegger had a number of other interactions and exchanged letters, and it is clear that Heidegger's philosophy was important to Celan. See James K. Lyon, *Martin Heidegger and Paul Celan: An Unresolved*

Conversation 1951–1970 (Baltimore: Johns Hopkins University Press, 2006). At the same time, Celan's poem "Todtnauberg" is a deeply ambivalent expression of his engagement with Heidegger, and he was repelled by Heidegger's political activities as much as he was fascinated by Heidegger's thought. I think that Derrida felt much the same way about Heidegger, and he identified with Celan's powerful attraction/repulsion.

8. The English translation is published in Jacques Derrida, *Sovereignties in Question: The Poetics of Paul Celan*, ed. Thomas Dutoit and Outi Pasanen (New York: Fordham University Press, 2005), 135–63.

9. See *Dialogue and Deconstruction: The Gadamer-Derrida Encounter*, ed. Diane P. Michelfelder and Richard E. Palmer (New York: SUNY Press, 1989), especially Part I.

10. Jacques Derrida, "Language Is Never Owned: An Interview," in *Sovereignties in Question*, 98.

11. Ibid.

12. Ibid., 99.

13. Jacques Derrida, "Shibboleth: For Paul Celan," in *Sovereignties in Question*, 23.

14. Ibid., 25.

15. Ibid., 59.

16. See *The Poetry of Paul Celan*, 339.

17. Derrida, "Shibboleth," 23.

18. See Jacques Derrida, "Violence and Metaphysics: An Essay on the Thought of Emmanuel Levinas," in *Writing and Difference*, trans. Alan Bass (Chicago: University of Chicago Press, 1978), 79–153.

19. Derrida, *Sovereignties in Question*, 140 (emphasis in original).

20. Ibid.

21. Ibid. Derrida also gives the original German: "Die Welt ist fort, ich muß dich tragen."

22. Hans-Georg Gadamer, *Gadamer on Celan: "Who Am I and Who Are You?" and Other Essays*, ed. Richard Heinemann and Bruce Krajewski (Albany: SUNY Press, 1997), 150.

23. Quoted in Derrida, *Sovereignties in Question*, 144. See also *Gadamer on Celan*, 95–96.

24. Derrida, *Sovereignties in Question*, 145.

25. Ibid., 147.

26. Ibid., 149.

27. Derrida, *Sovereignties in Question*, 153.

28. Ibid.

29. Ibid., 155.

30. Ibid.

31. Ibid., 157.

32. Ibid.

33. Ibid.

34. Xu Shen, *Shuo wen jie zi*, quoted in Li Zehou, *The Chinese Aesthetic Tradition*, trans. Maija Bell Samei (Manoa: University of Hawai'i Press, 2009), 1.

35. Li Zehou, *The Chinese Aesthetic Tradition*, 2.

36. I owe this specific formulation, as well as many conversations about Kierkegaard, Derrida, Lacan, Celan the *aqedah*, and especially the ram, to Timothy Snediker.

37. Derrida, *Sovereignties in Question*, 158.

38. Ibid. (emphasis in original).

39. Ibid., 159.

40. Martin Heidegger, *The Fundamental Concepts of Metaphysics: World, Finitude, Solitude*, trans. William McNeil and Nicholas Walker (Bloomington: Indiana University Press, 2001), 176–77.

41. An anecdote related by a colleague concerns Celan's visit to Germany in 1952 for a meeting of the poetry organization Gruppe 47 where a commotion broke out because a woman in a car had run over a dog. Celan remarked: "See how they carry on—about a dog." Quoted in Jerry Glenn, *Paul Celan* (New York: Twayne Publishers, 1973), 24. Celan was juxtaposing this concern for a dog with the lack of concern most Germans showed for Jews during the Nazi era.

42. Heidegger, *The Fundamental Concepts of Metaphysics*, 255.

43. Ibid.

44. Jacques Derrida, *The Animal That I Therefore Am*, trans. David Wills (New York: Fordham University Press, 2008), 142.

45. Heidegger, *The Fundamental Concepts of Metaphysics*, 154.

46. Derrida, *The Animal That I Therefore Am*, 155.

47. Heidegger, *The Fundamental Concepts of Metaphysics*, 196.

48. Derrida, *Sovereignties in Question*, 163.

49. Ibid.

50. Michael Naas, *Miracle and Machine: Jacques Derrida and the Two Sources of Religion, Science, and the Media* (New York: Fordham University Press, 2012), 95. One of the things that makes Naas's book so provocative is that he reads Derrida's essay "Faith and Knowledge" along with a novel by Don DeLillo, *Underworld*. Although I am unable to undertake it here, an interesting project that would be somewhat analogous would be to read Celan's speech on "The Meridian" along with Cormac McCarthy's novel *Blood Meridian: The Evening Redness in the West*.

51. Emmanuel Levinas, "Paul Celan: From Being to the Other," in *Proper Names*, trans. Michael B. Smith (Stanford: Stanford University Press, 1996), 46 (emphasis in original).

52. Ibid., 45 (emphasis in original).

53. *The Poetry of Paul Celan*, 211.

54. Jacques Derrida, *The Beast and the Sovereign*, Volume II, trans. Geoffrey Bennington (Chicago: University of Chicago Press, 2011), 9, 266.

55. Jacques Lacan, *On the Names-of-the-Father*, trans. Bruce Fink (Cambridge: Polity Press, 2013), 86.

56. Jacques Lacan, *Anxiety: The Seminar of Jacques Lacan Book X*, ed. Jacques-Alain Miller, trans. A. R. Price (Cambridge: Polity Press, 2014), 245.

57. Ibid., 307.

58. Lacan, *On the Names-of-the-Father*, 80.

59. Ibid., 87.

60. There is an interesting resemblance between the image of the ram "caught by its horns in a thicket" in Genesis 22:13, and the image of Absalom, David's rebellious son, who "was riding a mule and, as it passed beneath a great oak, his head was caught in its boughs" in 2 Samuel 18:9.

61. Lacan, *On the Names-of-the-Father*, 88.

62. Ibid., 88.

63. Ibid. On the topic of the *Aqedah* and its relation to child sacrifice more generally, see Jon D. Levenson, *The Death and Resurrection of the Beloved Son: The Transformation of Child Sacrifice in Judaism and Christianity* (New Haven: Yale University Press, 1993).

64. I am not going to contribute to this important literature and conversation here. For a couple of the many important discussions, see, in addition to Derrida's *The Animal That I Therefore Am*, Leonard Lawlor, *This is Not Sufficient: An Essay on Animality and Human Nature in Derrida* (New York: Columbia University Press, 2007), and Cary Wolfe, *Before the Law: Humans and Other Animals in a Biopolitical Frame* (Chicago: University of Chicago Press, 2012).

65. Derrida, *The Beast and the Sovereign*, Volume II, 6.

5. Derrida, Lacan, and Object-Oriented Ontology: Philosophy of Religion at the End of the World

1. See the recent volume edited by B. Keith Putt, Jeffrey W. Robbins, and myself, *The Future of Continental Philosophy of Religion* (Bloomington: Indiana University Press, 2014).

2. See Elizabeth Kolbert, *The Sixth Extinction: An Unnatural History* (New York: Henry Holt, 2014).

3. See Ray Brassier, Iain Hamilton Grant, Graham Harman, and Quentin Meillassoux, "Speculative Realism" in *Collapse: Philosophical Research and Development*, Volume III, 307–449; *The Speculative Turn: Continental Materialism and Realism*, ed. Levi Bryant, Nick Srnicek, and Graham Harman (Melbourne: re.press, 2011); and Graham Harman, *The Quadruple Object* (Alresford: Zero Books, 2010). I am not explicitly engaging with Harmon's philosophy in this chapter, but I do think that his work, while significant, also suffers from an oversimplified and idealized conceptualization of objects.

4. Quentin Meillassoux, *After Finitude: An Essay on the Necessity of Contingency*, trans. Ray Brassier (London: Continuum, 2008), 5.

5. Graham Harman, *Quentin Meillassoux: Philosophy in the Making* (Edinburgh: Edinburgh University Press, 2011), 80, 166.

6. See the counter-argument by Steven Shakespeare, "The Persistence of the Trace: Interrogating the Gods of Speculative Realism," in *The Future of*

Continental Philosophy of Religion, ed. Clayton Crockett, B. Keith Putt, and Jeffrey W. Robbins (Bloomington: Indiana University Press, 2014), 80–91.

7. Meillassoux, *After Finitude*, 10.

8. Ibid., 52.

9. Ibid., 60.

10. David Hume, *Dialogues Concerning Natural Religion*, ed. Martin Bell (New York: Penguin Books, 1990), 94–95.

11. Meillassoux, *After Finitude*, 104.

12. Ibid., 111.

13. Immanuel Kant, *Critique of Pure Reason*, trans. Norman Kemp Smith (New York: St. Martin's Press, 1965), 29.

14. Meillassoux, *After Finitude*, 46.

15. Ibid., 47.

16. Shakespeare, "The Persistence of the Trace," 83.

17. Excerpts from the unpublished French manuscript *L'Inexistence absolu* have been translated and published by Graham Harmon as an Appendix to *Quentin Meillassoux*, 175–238, quote 187.

18. Ibid., 238.

19. See Christopher Watkin, *Difficult Atheism: Post-Theological Thinking in Alain Badiou, Jean-Luc Nancy, and Quentin Meillassoux* (Edinburgh: Edinburgh University Press, 2013).

20. On this Hegelian distinction between *Vorstellung* and *Begriff*, as well as a critique of it, see John D. Caputo, *The Insistence of God: A Theology of Perhaps* (Bloomington: Indiana University Press, 2013), 88–97.

21. Jonathan Z. Smith, *Imagining Religion: From Babylon to Jonestown* (Chicago: University of Chicago Press, 1982), xi.

22. See José Casanova, *Public Religions in the Modern World* (Chicago: University of Chicago Press, 1994).

23. Talal Asad, *Formations of the Secular: Christianity, Islam, Modernity* (Stanford: Stanford University Press 2003), 200.

24. Jacques Derrida, *Rogues: Two Essays on Reason*, trans. Pascale-Anne Brault and Michael Naas (Stanford: Stanford University Press, 2005), 34.

25. Martin Heidegger, *The Fundamental Concepts of Metaphysics: World, Finitude, Solitude*, trans. William McNeil and Nicholas Walker (Bloomington: Indiana University Press, 2001), 176–77.

26. Martin Heidegger, "The Origin of the Work of Art," in Martin Heidegger, *Basic Writings*, ed. David Farrell Krell (New York: HarperCollins, 1993), 170.

27. Ibid.

28. Ibid.

29. Timothy Morton, *Hyperobjects: Philosophy and Ecology After the End of the World* (Minneapolis: University of Minnesota Press, 2013), 1.

30. Ibid., 7.

31. Ibid., 7 (emphasis in original).

32. Ibid., 128.

33. Ibid., 129.

34. Jacques Derrida, "Rams: Uninterrupted Dialogue—Between Two Infinities, the Poem," trans. Thomas Dutoit and Philippe Romanski, in Jacques Derrida, *Sovereignties in Question: The Poetics of Paul Celan*, ed. Thomas Dutoit and Outi Pasanen (New York: Fordham University Press, 2005), 163.

35. Ibid., 158 (emphasis Derrida's).

36. Jeffrey Jerome Cohen, *Stone: An Ecology of the Inhuman* (Minneapolis: University of Minnesota Press, 2015), 50, 43.

37. Morton, *Hyperobjects*, 136.

38. Jacques Derrida, *The Gift of Death*, trans. David Wills (Chicago: University of Chicago Press, 1995), 68.

39. Derrida, "Rams," 140.

40. Ibid.

41. Ibid.

42. Sam Weber, "Toward a Politics of Singularity," in *Crediting God: Sovereignty and Religion in the Age of Global Capitalism*, ed. Miguel Vater (New York: Fordham University Press, 2011), 237.

43. Ibid., 240.

44. Ibid., 246.

45. Derrida, *The Gift of Death*, 87.

46. Morton, *Hyperobjects*, 70.

47. Morton, *Hyperobjects*, 147.

48. Ibid., 148.

49. This is a play on words, where the French phrase sounds just like Lacan's phrase, "*le nom du pere*," or the Name of the Father, which guarantees the symbolic order. See Lacan, Seminar XXI, 1973–1974, *Les non-dupes errant*. This seminar has not been published in English, but an English translation by Cormac Gallagher can be accessed online at Lacanian Works: http://www .lacanianworks.net/?p=807.

50. Michael S. Northcott, *A Political Theology of Climate Change* (Grand Rapids, Mich.: William B. Eerdmann's Publishing Co., 2013), 6.

51. See Clayton Crockett and Jeffrey W. Robbins, *Religion, Politics and the Earth: The New Materialism* (New York: Palgrave Macmillan, 2012).

52. Ibid., 72.

53. For a fuller treatment of Deleuze, focusing on these two works, see my book *Deleuze Beyond Badiou: Ontology, Multiplicity and Event* (New York: Columbia University Press, 2013).

54. See my chapter on "Entropy" in *The Future of Continental Philosophy of Religion*, 272–87.

6. Radical Theology and the Event: Caputo's Derridean Gospel

1. Mark C. Taylor, *Erring: A Postmodern A/theology* (Chicago: University of Chicago Press, 1984), 10.

2. Ibid., 11 (emphasis Taylor's).

3. Charles E. Winquist, *Epiphanies of Darkness: Deconstruction in Theology* (Aurora, Colo.: The Davies Group, Publishers, 1998 [1986]), 53.

4. Carl A. Raschke, "The Deconstruction of God," in Thomas J. J. Altizer, et. al., *Deconstruction & Theology* (New York: Crossroad, 1982), 1–33 (quote 3, emphasis Raschke's).

5. Mark C. Taylor, *Hiding* (Chicago: University of Chicago Press, 1997), 272.

6. Ibid.

7. Charles E. Winquist, "Postmodern Secular Theology," in *Secular Theology: American Radical Theological Thought*, ed. Clayton Crockett (London: Routledge, 2001), 26–36 (quote 31).

8. Ibid.

9. Ibid., 36.

10. Ibid., 31.

11. John D. Caputo, *Radical Hermeneutics: Repetition, Deconstruction, and the Hermeneutic Project* (Bloomington: Indiana University Press, 1987), 5.

12. Ibid.

13. Ibid., 192.

14. Ibid., 200.

15. Ibid., 272.

16. Ibid., 280.

17. Ibid., 289.

18. John D. Caputo, *The Prayers and Tears of Jacques Derrida: Religion Without Religion* (Bloomington: Indiana University Press, 1997), xix (emphasis Caputo's). The word *pas* in French means both "step" and "not," and Derrida and Blanchot both play on this double meaning.

19. Ibid., xx.

20. See ibid., 2–3.

21. Jacques Derrida, *The Gift of Death*, trans. David Wills (Chicago: University of Chicago Press, 1995), 49 (emphasis Derrida's).

22. Caputo, *Prayers and Tears*, xxi.

23. Ibid., 16 (emphasis Caputo's).

24. Ibid.

25. Ibid., 195.

26. Ibid., 233.

27. Ibid., 289.

28. Ibid., 61.

29. Ibid., 62.

30. John D. Caputo, *The Weakness of God: A Theology of the Event* (Bloomington: Indiana University Press, 2006), 2.

31. Ibid. (emphasis Caputo's).

32. Ibid., 3.

33. Ibid., 6.

34. Ibid., 9.

35. See Jean-Luc Marion, *God Without Being: Hors-Texte*, trans. Thomas A. Carlson (Chicago: University of Chicago Press, 1991.

36. Caputo, *The Weakness of God*, 39 (emphasis Caputo's).

37. Ibid., 240.

38. See ibid., 109: "This is why, beyond any Greek sense of wonder, the texts of the kingdom read—if we may adopt a suggestion coming from Gilles Deleuze—like a veritable Alice in Wonderland of wedding feasts as mad as any hatter's party, of sinners getting preference over perfectly respectable fellows, of virgins giving birth, of eventualities that confound the economy of the world."

39. Ibid., 238.

40. Ibid.

41. Ibid., 252.

42. Ibid.

43. Ibid., 247.

44. John D. Caputo, *The Insistence of God: A Theology of Perhaps* (Blooming-ton: Indiana University Press, 2013), 15.

45. Ibid., 19 (emphasis Caputo's).

46. Ibid., 45.

47. Ibid., 68.

48. Ibid., 69.

49. Ibid.

50. Ibid., 100.

51. Ibid., 101.

52. Ibid., 92.

53. Ibid., 123.

54. Ibid., 125.

55. Ibid.

56. See Katrin Pahl, *Tropes of Transport: Hegel and Emotion* (Evanston: Northwestern University Press, 2012).

57. See Catherine Malabou, *The Future of Hegel: Plasticity, Temporality and Dialectic*, trans. Lisbeth During (London: Routledge, 2004).

58. Caputo, *The Insistence of God*, 131.

59. Jacques Derrida, "A Time for Farewells," in Malabou, *The Future of Hegel*, xlvii. Malabou addresses this possibility of the absolute accident in Catherine Malabou, *Ontology of the Accident: An Essay on Destructive Plasticity*, trans. Carolyn Shread (Cambridge: Polity Press, 2012).

60. Ibid., 126.

61. Ibid., 133.

7. Deconstructive Plasticity: Malabou's Biological Materialism

1. See Catherine Malabou, *The Future of Hegel: Plasticity, Temporality, and Dialectic*, trans. Lisbeth Durling (London: Routledge, 2004).

2. Catherine Malabou, *Plasticity at the Dusk of Writing: Dialectic, Destruction, Deconstruction*, trans. Carolyn Shread (New York: Columbia University Press, 2009), 12 (emphasis Malabou's).

3. Ibid. (emphasis Malabou's).

4. Ibid., 15 (emphasis Malabou's).

5. Ibid., 57.

6. Ibid.

7. Ibid., 60 (emphasis Malabou's).

8. Catherine Malabou, *Changing Difference: The Feminine and the Question of Philosophy*, trans. Carolyn Shread (London: Polity, 2009), 48.

9. Ibid., 55.

10. Malabou, *Plasticity at the Dusk of Writing*, 57 (emphasis Malabou's).

11. See Alain Badiou, *The Century*, trans. Alberto Toscano (London: Polity, 2007).

12. Malabou, *Changing Difference*, 66.

13. Catherine Malabou, *What Should We Do with Our Brain?*, trans. Marc Jeannerod (New York: Fordham University Press, 2008), 5.

14. Ibid., 74.

15. Catherine Malabou, *The New Wounded: From Neurosis to Brain Damage*, trans. Steven Miller (New York: Fordham University Press, 2012), 141 (emphasis Malabou's).

16. Ibid., 165.

17. Adrian Johnston and Catherine Malabou, *Self and Emotional Life* (New York: Columbia University Press, 2013), 11.

18. Ibid.

19. Ibid., 33 (emphasis Malabou's).

20. Ibid., 58.

21. Ibid.

22. Malabou, *Changing Difference*, 73.

23. Ibid., 81.

24. Malabou, *What Should We Do with Our Brain?*, 73.

25. Ibid., 74.

26. Malabou, *The New Wounded*, 141 (emphasis Malabou's).

27. Malabou, *Changing Difference*, 82.

28. Ibid., 83.

29. Ibid., 87 (emphasis Malabou's).

30. Ibid.

31. Ibid.

32. Johnston and Malabou, *Self and Emotional Life*, 72.

33. See Edward Baring, *The Young Derrida and French Philosophy, 1945–1968* (Cambridge: Cambridge University Press, 2011), Chapter 5, 146–81.

34. Malabou, *What Should We Do with Our Brain?*, 38.

35. Ibid.

36. Malabou, *Plasticity at the Dusk of Writing*, 61 (emphasis Malabou's).

37. Catherine Malabou, "Darwin and the Social Destiny of Nature," in "Plastique: The Dynamics of Catherine Malabou," *theory@buffalo* 16, ed. Jarrod Abbott and Tyler Williams, 2012, 144–56 (quote 144).

38. Ibid., 145.

39. Ibid., 152.

40. Ibid., 153.

41. Ibid.

42. See Eva Jablonka and Marion J. Lamb, *Evolution in Four Dimensions: Genetic, Epigenetic, Behavioral, and Symbolic Variations in the History of Life* (Cambridge: MIT Press, 2005).

43. Malabou, "Darwin and the Social Destiny of Natural Selection," 155.

44. Catherine Malabou, "The Future of Derrida," in *The Future of Continental Philosophy of Religion*, ed. Clayton Crockett, B. Keith Putt, and Jeffrey W. Robbins (Bloomington: Indiana University Press, 2014), 209–18 (quote 210). See also Catherine Malabou, *Before Tomorrow: Epigenesis and Rationality*, trans. Carolyn Shread (Cambridge, UK: Polity Press, 2016).

45. Ibid., 211.

46. Ibid.

47. Ibid., 215.

48. Ibid., 216.

49. Ibid., 217.

8. Quantum Derrida: Barad's Hauntological Materialism

1. Ibid., 215.

2. Ibid.

3. Jacques Derrida, *Of Spirit: Heidegger and the Question*, trans. Geoffrey Bennington and Rachel Bowlby (Chicago: University of Chicago Press, 1989), 47.

4. See Jacques Derrida, *The Beast and the Sovereign* Volume II, trans. Geoffrey Bennington (Chicago: University of Chicago Press, 2011), 6.

5. Jacques Derrida, *Sovereignties in Question: The Poetics of Paul Celan*, ed. Thomas Dutoit and Outi Pasanen (New York: Fordham University Press, 2005), 163.

6. Jacques Derrida, *The Gift of Death*, trans. David Wills (Chicago: University of Chicago Press, 1995), 86.

7. Ibid.

8. Ibid., 87.

9. Quoted by Derrida in ibid., 98.

10. Ibid., 106.

11. Ibid., 114.

12. Ibid., 115.

13. Ibid.

14. Thomas S. Kuhn, *The Structure of Scientific Revolutions*, fourth edition (Chicago: University of Chicago Press, 2012), 157 (emphasis mine).

15. Friedrich Nietzsche, *On the Genealogy of Morals and Ecce Homo*, trans. Walter Kaufmann and R. J. Hollingdale (New York: Random House, 1967), 155.

16. Karen Barad, *Meeting the Universe Halfway: Quantum Physics and the Entanglement of Matter and Meaning* (Durham: Duke University Press, 2007), 85.

17. Ibid., 141.

18. Ibid., 140.

19. Ibid., 141.

20. Ibid., 333.

21. McKenzie Wark, *Molecular Red: Theory for the Anthropocene* (London: Verso, 2015), 158.

22. Barad, *Meeting the Universe Halfway*, 102.

23. Ibid., 106.

24. Ibid., 307.

25. Ibid., 309.

26. Karen Barad, "Quantum Entanglements and Hauntological Relations of Inheritance: Dis/continuities, SpaceTime Enfoldings, and Justice-to-Come," *Derrida Today* 3.2, 2010, 240–68 (quote 251).

27. Ibid., 251.

28. Ibid., 264 (emphasis in original).

29. Ibid.

30. Karen Barad, "On Touching: The Inhuman That I Therefore Am," *Differences: A Journal of Feminist Cultural Studies* 25, no. 3 (2012), 206–23 (quote 209).

31. Ibid., 210.

32. Ibid.

33. Quoted in ibid., 212.

34. Ibid., 210 (emphasis in original).

35. Ibid., 217.

36. Ibid., 209.

37. See the marvelous book on cosmology by Mary-Jane Rubenstein, *Worlds Without End: The Many Lives of the Multiverse* (New York: Columbia University Press, 2014).

38. Barad, "The Inhuman That I Therefore Am," 218 (emphasis in original).

39. François Laruelle, *Philosophie Non-Standard: Générique, Quantique, Philo-Fiction* (Paris: Éditions Kimé, 2010), 13, which I have paraphrased: "*Reste que la non-philosophie ne prétend surtout pas intervenir directement dans la physique quantique, mais se tenir en revanche à la philosophie dans le même rapport que la physique quantique à la physique classique, rapport qui est moin d'englobant que d'une certaine généralization.*"

40. Ibid., 125.

41. See ibid., 142.

42. Jacques Derrida and François Laruelle, "Controversy over the Possibility of a Science of Philosophy," trans. Ray Brassier and Robin Mackay, in *The Non*

Philosophy Project: Essays by François Laruelle, ed. Gabriel Alkon and Boris Gunjevic (New York: Telos Press, 2010), 77.

43. Ibid., 89.

44. Gilles Deleuze and Félix Guattari, *What Is Philosophy?*, trans. Hugh Tomlinson and Graham Burchell (New York: Columbia University Press, 1994), 220.

45. Ibid., 234.

46. Barad, *Meeting the Universe Halfway*, 72.

47. Ibid., 73.

48. Ibid., 76.

49. Ward, *Molecular Red*, 154.

50. Ibid., 154.

51. Gilles Deleuze, *Difference and Repetition*, trans. Paul Patton (New York: Columbia University Press, 1994), 118. See also my book *Deleuze Beyond Badiou: Ontology, Multiplicity and Event* (New York: Columbia University Press. 2013), Chapter 3, which also considers this problem of the relation of difference to difference in *Difference and Repetition*, although it does not use the term diffraction pattern.

52. Ibid., 119.

53. Ibid.

54. Ibid.

55. Ibid.

56. Ibid., 120.

57. Jacques Derrida, "*Différance*," in Jacques Derrida, *Margins of Philosophy*, trans. Alan Bass (Chicago: University of Chicago Press, 1982), 3–27, quote 8.

58. Ibid., 8.

59. Ibid., 13.

Afterword: The Sins of the Fathers—A Love Letter

1. See Jacques Lacan, *Transference: The Seminar of Jacques Lacan Book VIII*, ed. Jacques-Alain Miller, trans. Bruce Fink (Cambridge: Polity Press, 2015).

2. I am aware that I am white, and that whiteness deforms and distorts the social symbolic field with a force all its own. This introduction is limited insofar as I am not explicitly reflecting on race personally or professionally, and specifically not in connection to Derrida's philosophy. This is a lack, but it is not something I am able to address here. In psychoanalysis, gender and sexuality have been foregrounded, but race has not been nearly as significant. Judith Butler has done a great deal to show how race, gender, and sexuality are mutually configured—see, for instance, Chapter 6 of *Bodies That Matter: On the Discursive Limits of "Sex"* (London: Routledge, 1993). The analyses of whiteness in contemporary critical race theory are absolutely crucial, and need to be further integrated with continental philosophy: See the pioneering work of George Yancy, *Black Bodies, White Gazes: The Continuing Significance of Race* (Lanham, Md.: Rowman and Littlefield, 2008). See also the important work of Linda Martín

Alcoff, *Visible Identities: Race, Gender and the Self* (Oxford: Oxford University Press, 2005), and *The Future of Whiteness* (Cambridge: Polity Press, 2015).

3. Sigmund Freud, *Civilization and Its Discontents*, trans. James Strachey (New York: Norton, 1961), 20.

4. Quoted in Geoffrey Bennington, *No Half No End: Militantly Melancholic Essays in Memory of Jacques Derrida* (Edinburgh: Edinburgh University Press, 2010), 15.

5. Julia Kristeva, *Revolution in Poetic Language*, trans. Margaret Walker (New York: Columbia University Press, 1984), 25.

6. Ibid., 69.

7. Julia Kristeva, *This Incredible Need to Believe*, trans. Beverly Bie Brahic (New York: Columbia University Press, 2009), xi.

8. See Benoît Peeters, *Derrida: A Biography*, trans. Andrew Brown (Cambridge: Polity Press, 2013), 356.

9. Jacques Derrida, *Points . . . Interviews, 1974–1994*, ed. Elisabeth Weber, trans. Peggy Kamuf & Others (Stanford: Stanford University Press, 1995), 199.

10. Ibid.

11. Noëlle Vahanian, *The Rebellious No: Variations on a Secular Theology of Language* (New York: Fordham University Press, 2014), 15.

12. Ibid.

13. Catherine Malabou, *Changing Difference: The Feminine and the Question of Philosophy*, trans. Carolyn Shread (Cambridge, UK: Polity Press, 2011), 138.

14. Ibid., 34.

15. Ibid., 140.

16. Ibid., 35.

17. Ibid., 36.

18. Ibid., 37.

19. Ibid., 39.

20. Ibid., 39–40.

21. Ibid., 140–41.

22. Ibid., 141.

23. Bracha L. Ettinger, *The Matrixial Borderspace* (Minneapolis: University of Minnesota Press, 2006), 64 (emphasis in original). This matrixial borderspace also resonates in artistic and aesthetic terms with the "borderlands" articulated by Gloria Anzaldúa. See Gloria Anzaldúa, *Borderlands/La Frontera: The New Mestiza* (San Francisco: Aunt Lute Books, 2012).

24. Bracha L. Ettinger, "Copoeisis," in *Ephemera: Theory & Politics in Organization*, Vol. 5(X), 703–13, quote 704.

25. Catherine Keller, *Face of the Deep: A Theology of Becoming* (London: Routledge, 2003), 165 (emphasis in original).

26. Catherine Keller, *Cloud of the Impossible: Negative Theology and Planetary Entanglement* (New York: Columbia University Press, 2016), 314.

27. Jacques Derrida, *Margins of Philosophy*, trans. Alan Bass (Chicago: University of Chicago Press, 1982), 135.

28. Ibid.

29. Ibid., 135–36.

30. Keller, *Cloud of the Impossible*, 282.

31. Ibid.

32. Friedrich Nietzsche, *The Portable Nietzsche*, ed. and trans. Walter Kaufmann (New York: Penguin Books, 1954), 437.

33. Ibid. (emphasis mine).

34. Ibid., 438 (emphasis in original).

35. See Keller, *Face of the Deep*, 29: "to love is to bear with the chaos."

36. Nietzsche, *The Portable Nietzsche*, 439 (emphasis in original).

37. Colleen Hartung, "Faith and Polydoxy in the Whirlwind," in *Polydoxy: Theology of Multiplicity and Relation*, ed. Catherine Keller and Laurel C. Schneider (New York: Routledge, 2011), 152. See also my response to the volume *Polydoxy* in a Special Issue of the journal *Modern Theology*, guest edited by Mary-Jane Rubenstein and Kathryn Tanner, from which this discussion partly draws: Clayton Crockett, "Polyhairesis: On Postmodern and Chinese Folds," *Modern Theology* 30, no. 3 (2014), 34–49.

38. Ibid., 153.

39. Ibid., 154.

40. Ibid., 155.

41. Ibid., 158.

42. Ibid., 159.

43. Ibid.

44. Ibid., 160.

45. Ibid., 161.

46. Ibid., 162.

47. Ibid.

48. Ibid., 163.

49. Bennington, *Not Half No End*, 9.

50. Ibid., xi.

51. Katerina Kolozova, *Cut of the Real: Subjectivity in Poststructuralist Philosophy* (New York: Columbia University Press, 2014), 106.

52. Ibid., 108.

53. Ibid., 113.

54. Ibid., 118.

55. Ibid.

56. Ibid., 124.

57. Ibid., 125.

58. Ibid., 128.

59. Ibid., 129.

60. Ibid.

Index

Perspectives in Continental Philosophy
John D. Caputo, series editor

Recent titles:

Richard Kearney and Brian Treanor, eds., *Carnal Hermeneutics*.

Aaron T. Looney, *Vladimir Jankélévitch: The Time of Forgiveness*.

Vanessa Lemm, ed., *Nietzsche and the Becoming of Life*.

Edward Baring and Peter E. Gordon, eds., *The Trace of God: Derrida and Religion*.

Jean-Louis Chrétien, *Under the Gaze of the Bible*. Translated by John Marson Dunaway.

Michael Naas, *The End of the World and Other Teachable Moments: Jacques Derrida's Final Seminar*.

Noëlle Vahanian, *The Rebellious No: Variations on a Secular Theology of Language*.

A complete list of titles is available at http://fordhampress.com.